T0281374

Docker for Data Science

Building Scalable and Extensible Data Infrastructure Around the Jupyter Notebook Server

Joshua Cook

Apress®

Docker for Data Science

Joshua Cook
Santa Monica, California, USA

ISBN-13 (pbk): 978-1-4842-3011-4   ISBN-13 (electronic): 978-1-4842-3012-1
DOI 10.1007/978-1-4842-3012-1

Library of Congress Control Number: 2017952396

Copyright © 2017 by Joshua Cook

This work is subject to copyright. All rights are reserved by the Publisher, whether the whole or part of the material is concerned, specifically the rights of translation, reprinting, reuse of illustrations, recitation, broadcasting, reproduction on microfilms or in any other physical way, and transmission or information storage and retrieval, electronic adaptation, computer software, or by similar or dissimilar methodology now known or hereafter developed.

Trademarked names, logos, and images may appear in this book. Rather than use a trademark symbol with every occurrence of a trademarked name, logo, or image we use the names, logos, and images only in an editorial fashion and to the benefit of the trademark owner, with no intention of infringement of the trademark.

The use in this publication of trade names, trademarks, service marks, and similar terms, even if they are not identified as such, is not to be taken as an expression of opinion as to whether or not they are subject to proprietary rights.

While the advice and information in this book are believed to be true and accurate at the date of publication, neither the authors nor the editors nor the publisher can accept any legal responsibility for any errors or omissions that may be made. The publisher makes no warranty, express or implied, with respect to the material contained herein.

Cover image by Freepik (`www.freepik.com`)

Managing Director: Welmoed Spahr
Editorial Director: Todd Green
Acquisitions Editor: Celestin Suresh John
Development Editor: Laura Berendson
Technical Reviewer: Jeeva S. Chelladhurai
Coordinating Editor: Prachi Mehta
Copy Editor: Mary Behr

Distributed to the book trade worldwide by Springer Science+Business Media New York, 233 Spring Street, 6th Floor, New York, NY 10013. Phone 1-800-SPRINGER, fax (201) 348-4505, e-mail `orders-ny@springer-sbm.com`, or visit `www.springeronline.com`. Apress Media, LLC is a California LLC and the sole member (owner) is Springer Science + Business Media Finance Inc (SSBM Finance Inc). SSBM Finance Inc is a **Delaware** corporation.

For information on translations, please e-mail `rights@apress.com`, or visit `www.apress.com/rights-permissions`.

Apress titles may be purchased in bulk for academic, corporate, or promotional use. eBook versions and licenses are also available for most titles. For more information, reference our Print and eBook Bulk Sales web page at `www.apress.com/bulk-sales`.

Any source code or other supplementary material referenced by the author in this book is available to readers on GitHub via the book's product page, located at `www.apress.com/978-1-4842-3011-4`. For more detailed information, please visit `www.apress.com/source-code`.

Printed on acid-free paper

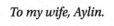

To my wife, Aylin.

Contents at a Glance

Contents

About the Author

Joshua Cook is a mathematician. He writes code in Bash, C, and Python and has done pure and applied computational work in geo-spatial predictive modeling, quantum mechanics, semantic search, and artificial intelligence. He also has 10 years experience teaching mathematics at the secondary and post-secondary level. His research interests lie in high-performance computing, interactive computing, feature extraction, and reinforcement learning. He is always willing to discuss orthogonality or to explain why Fortran is the language of the future over a warm or cold beverage.

About the Technical Reviewer

Jeeva S. Chelladhurai has been working as a DevOps specialist at the IBM GTS Labs for the last 9 years. He is the co-author of *Learning Docker*, published by PacktPub, UK. He has more than 20 years of IT industry experience. He has technically managed and mentored diverse teams across the globe in envisaging and building pioneering telecommunication products. He specializes in DevOps, automation, and cloud solution delivery, with a focus on data center optimization, software-defined environments (SDEs), and distributed application development, deployment, and delivery using the newest Docker technology. Jeeva is also a strong proponent of the agile methodologies, DevOps, and IT automation. He holds a master's degree in computer science from Manonmaniam Sundaranar University and a graduation certificate in project management from Boston University, Boston, Massachusetts, USA. Besides his official responsibilities, he writes book chapters and authors research papers. He has been instrumental in crafting reusable technical assets for IBM solution architects and consultants. He speaks in technical forums on DevOps technologies and tools. He hosts one of the largest Open Source communities in Bangalore (www.meetup.com/opensourceblr/). His LinkedIn profile can be found at www.linkedin.com/in/JeevaChelladhurai.

Acknowledgments

Thanks to Mike Frantz, Gilad Gressel, Devon Muraoka, Bharat Ramanathan, Nash Taylor, Matt Zhou, and DSI Santa Monica Cohorts 3 and 4 for talking through some of the more abstract concepts herein with me. Thanks to Chad Arnett for keeping it weird. Thanks to Jim Kruidenier and Jussi Eloranta for teaching an old dog new tricks. Thanks to my father for continually inspiring me and my mother for giving me my infallible belief in goodness. Thanks to Momlo for paving the way and Dablo for his curiosity. Thanks to my wife, Aylin, for her belief in me and tolerance for the word "eigenvector."

Introduction

This text is designed to teach the concepts and techniques of Docker and its ecosystem as applied to the field of data science. Besides introducing the core Docker technologies (the container and image, the engine, the Dockerfile), this book contains a discussion on building larger integrated systems using the Jupyter Notebook Server and open source data stores MongoDB, PostgreSQL, and Redis.

The first chapter walks the reader through a recommended hardware configuration for working through the text using an AWS t2.micro. Chapters 2 and 3 introduce the core technologies used in the book, Docker and Jupyter, as well as the idea of interactive programming. Chapters 4, 5, 6, and 9 dig deeper into specific areas of the Docker ecosystem. Chapter 7 explores the official Jupyter Docker images developed and maintained by the Jupyter development team. Chapter 8 introduces the Docker images for three open source data stores. Chapters 9 and 10 tie everything together, connecting Jupyter to data stores using Docker Compose. After having completed the book, readers are encouraged to reread Chapter 3 and Chapter 10 to begin to develop their own interactive software development style.

The concepts presented herein can be challenging, especially in terms of the abstraction of computer resources and processes. That said, no requisite knowledge is assumed. An attempt has been made to build the discussion from base principles. With this in mind, the reader should be comfortable working at the command line and have an adventurous and inquisitive spirit. We hope that readers with an intermediate to advanced understanding of Docker, Jupyter, or both will gain a deeper understanding of the concepts and learn novel approaches to the solving of computational problems using these tools.

CHAPTER 1

■ ■ ■

Introduction

The typical data scientist consistently has a series of extremely complicated problems on their mind beyond considerations stemming from their system infrastructure. Still, it is inevitable that infrastructure issues will present themselves. To oversimplify, we might draw a distinction between the "modeling problem" and the "engineering problem." The data scientist is uniquely qualified to solve the former, but can often come up short in solving the latter.

Docker has been widely adopted by the system administrator and DevOps community as a modern solution to the challenges presented in high availability and high performance computing.[1] Docker is being used for the following: transitioning legacy applications to a more modern "microservice"-based approach, facilitating continuous integration and continuous deployment for the integration of new code, and optimizing infrastructure usage.

In this book, I discuss Docker as a tool for the data scientist, in particular in conjunction with the popular interactive programming platform Jupyter. Using Docker and Jupyter, the data scientist can easily take ownership of their system configuration and maintenance, prototype easily deployable and scalable data solutions, and trivially clone entire systems with an eye toward replicability and communication. In short, I propose that skill with Docker is just enough knowledge of systems operations to make the data scientist dangerous. Having done this, I propose that Docker can add high performance and availability tools to the data scientist's toolbelt and fundamentally change the way that models are developed, prototyped, and scaled.

"Big Data"

A precise definition of "big data" will elude even the most seasoned data wizard. I favor the idea that big data is the exact scope of data that is no longer manageable without explicit consideration to its scope. This will no doubt vary from individual to individual and from development team to development team. I believe that mastering the concepts and techniques associated with Docker presented herein will drastically increase the size and scope of what exactly big data is for any reader.

---

[1] www.docker.com/use-cases

© Joshua Cook 2017
J. Cook, *Docker for Data Science*, DOI 10.1007/978-1-4842-3012-1_1

Recommended Practice for Learning

In this first chapter, you jump will headlong into using Docker and Jupyter on a cloud system. I hope that readers have a solid grasp of the Python numerical computing stack, although I believe that nearly anyone should be able to work their way through this book with enough curiosity and liberal Googling.

For the purposes of working through this book, I recommend using a sandbox system. If you are able to install Docker in an isolated, non-mission critical setting, you can work through this text without fear of "breaking things." For this purpose, I here describe the process of setting up a minimal cloud-based system for running Docker using Amazon Web Services (AWS).

As of the writing of this book, AWS is the dominant cloud-based service provider. I don't endorse the idea that its dominance is a reason a priori to use its services. Rather, I present an AWS solution here as one that will be the easiest to adopt by the largest group of people. Furthermore, I believe that this method will generalize to other cloud-based offerings such as DigitalOcean[2] or Google Cloud Platform,[3] provided that the reader has secure shell (ssh) access to these systems and that they are running a Linux variant.

I present instructions for configuring a system using Elastic Compute Cloud (EC2). New users receive 750 hours of free usage on their T2.micro platform and I believe that this should be more than enough for the typical reader's journey through this text.

Over the next few pages, I outline the process of configuring an AWS EC2 system for the purposes of working through this text. This process consists of

1. Configuring a key pair

2. Creating a new security group

3. Creating a new EC2 instance

4. Configuring the new instance to use Docker

Set up a New AWS Account

To begin, set up an AWS account if you do not already have one.[4]

■ **Note**　This work can be done in any region, although it is recommended that readers take note of which region they have selected for work (Figure 1-1). For reasons I have long forgotten, I choose to work in us-west-2.

---

[2]www.digitalocean.com
[3]https://cloud.google.com
[4]Instructions for creating a new AWS account can be found at https://aws.amazon.com/premiumsupport/knowledge-center/create-and-activate-aws-account/.

Figure 1-1. *Readers should take note of the region in which they are working*

Configure a Key Pair

In order to interface with your sandbox system running on AWS EC2, you will need an ssh key pair. Amazon EC2 uses public-key cryptography to facilitate all connections to running EC2 instances.[5] In your case, this means the creation of a secure connection between your local system and a sandbox system you will configure on an EC2 instance. To do this, you will create an ssh key pair locally and import the public component of the key pair into AWS. When you create a new instance, you have AWS provision the new instance with the public key, so that you can use your local private key to connect to the instance.

■ **Note**   Windows users are encouraged to make use of the excellent Git BASH tool available as part of the Git for Windows package here: `https://git-for-windows.github.io`. Git BASH will include all of the necessary command line tools that you will be using, including `ssh-keygen` and `ssh`.

In Listing 1-1, you use the `ssh-keygen` tool to create a key pair on your local system. For these purposes (that is, a disposable sandbox AWS system), you can leave all fields blank, including a passphrase to use the ssh key. The location in which you save the key will vary from system to system. The default location on my system is `~/.ssh/id_rsa` where ~ signifies the current user's home directory.[6] This process will create `id_rsa` and `id_rsa.pub`, a key pair. You will place the `id_rsa.pub` key into systems you wish to access and thus be able to ssh into these systems using the `id_rsa` file.

Listing 1-1. Create a New Key Pair

```
$ ssh-keygen -t rsa
Generating public/private rsa key pair.
Enter file in which to save the key (/home/ubuntu/.ssh/id_rsa):
Enter passphrase (empty for no passphrase):
Enter same passphrase again:
```

---

[5]`http://docs.aws.amazon.com/AWSEC2/latest/UserGuide/ec2-key-pairs.html`
[6]`www.gnu.org/software/bash/manual/html_node/Tilde-Expansion.html`

Your identification has been saved in /home/ubuntu/.ssh/id_rsa.
Your public key has been saved in /home/ubuntu/.ssh/id_rsa.pub.
The key fingerprint is:
SHA256:g5IYNQMf1n1jW5p36Y9I/qSPxnckhT665KtiBO6xu2U ubuntu@ip-172-31-43-19
The key's randomart image is:
```
+---[RSA 2048]----+
|   ..*. .        |
|   + +. . + .    |
|   .. o * o |
|   o . .. + . + .|
|   . o . So . + . |
|   . +. . = .|
|   o oE+.o.* |
|   =o.o*+o o|
|   ..+.o**o. |
+----[SHA256]-----+
```

In Listing 1-2, you verify that the contents of the key using the cat tool. You display a public key that was created on a remote Ubuntu system, as can be seen at the end of the key (ubuntu@ip-172-31-43-19). This should appear similar on any system.

Listing 1-2. Verify Newly Created ssh-key

```
$ cat ~/.ssh/id_rsa.pub
ssh-rsa
AAAAB3NzaC1yc2EAAAADAQABAAAABAQDdnHPEiq1a4OsDDY+g9luWQS8pCjBmR
64MmsrQ9MaIaE5shIcFB1Kg3pGwJpypiZjoSh9pS55S9LckNsBfn8Ff42ALLj
R8y+WlJKVk/ODvDXgGVcCcOt/uTvxVxObRruYxLW167J89UnxnJuRZDLeY9fD
OfIzSR5eglhCWVqiOzB+OsLqR1WO4Xz1oStID78UiY5msW+EFg25Hg1wepYMC
JG/Zr43ByOYPGseUrbCqFBS1KlQnzfWRfEKHZbtEe6HbWwz1UDL2NrdFXxZAI
XYYoCVtl4WXd/WjDwSjbMmtf3BqenVKZcP2DQ9/W+geIGGjvOTfUdsCHennYI
EUfEEP ubuntu@ip-172-31-43-19
```

Create a New Key Pair on AWS

Log in to your AWS control panel and navigate to the EC2 Dashboard, as shown in Figure 1-2. First, access "Services" (Figure 1-2, #1) then access "EC2" (Figure 1-2, #2). The Services link can be accessed from any page in the AWS website.

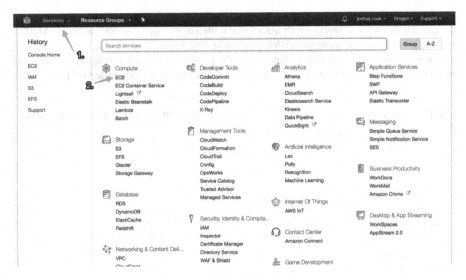

Figure 1-2. *Access the EC2 control panel*

Once at the EC2 control panel, access the Key Pairs pane using **either link** (Figure 1-3).

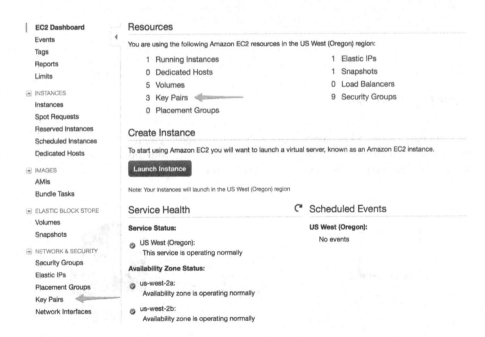

Figure 1-3. *Access key pairs in the EC2 Ddashboard*

From the Key Pairs pane, choose "Import Key Pair." This will activate a modal that you can use to create a new key pair associated with a region on your AWS account. Make sure to give the key pair a computer-friendly name, like from-MacBook-2017. Paste the contents of your public key (id_rsa.pub) into the public key contents. Prior to clicking Import, your key should appear as in Figure 1-4. Click Import to create the new key.

■ **Note** Many AWS assets are created uniquely by region. Key pairs created in one region will not be available in another.

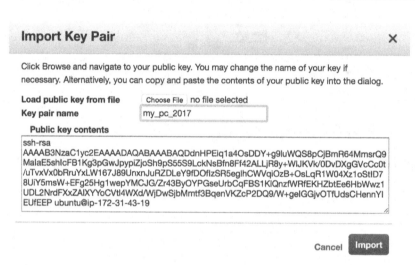

Figure 1-4. Import a new key pair

You have created a key pair between AWS and your local system. When you create a new instance, you will instruct AWS to provision the instance with this key pair and thus you will be able to access the cloud-based system from your local system.

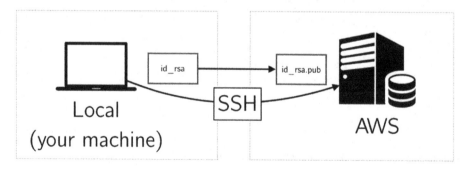

Figure 1-5. Connect to AWS from your local machine using an SSH key

▨ **Note** The terminology can be a bit confusing. AWS refers to an uploaded **public** key as a "key pair." To be clear, you are uploading the public component of a key pair you have created on your system (e.g. id_rsa.pub). The **private** key will remain on your system (e.g. id_rsa).

Create a New Security Group

From the EC2 Dashboard, access the Security Group pane using **either link** (Figure 1-6).

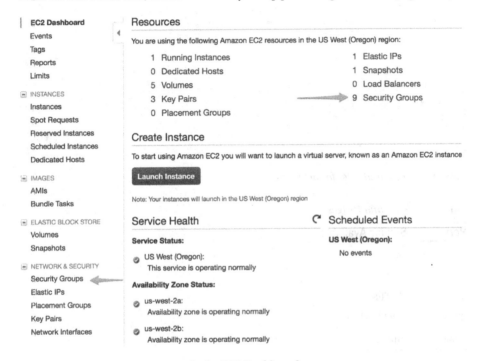

Figure 1-6. *Access security groups in the EC2 Dashboard*

From the Security Group pane, click "Create Security Group." Give the security group a computer friendly group name like jupyter_docker. Give the security group a description like "Open access to Jupyter and Docker default ports." Use the default VPC. Access the Inbound tab and configure the following security rules:

- SSH: Port Range: 22, Source: Anywhere
- HTTP: Port Range: 80, Source: Anywhere
- HTTPS: Port Range: 443, Source: Anywhere
- Custom TCP Rule: Port Range: 2376, Source: Anywhere
- Custom TCP Rule: Port Range: 8888, Source: Anywhere

When you have added all of these rules, it should appear as in Figure 1-7. Table 1-1 shows a list of ports and the services that will be accessible over these ports.

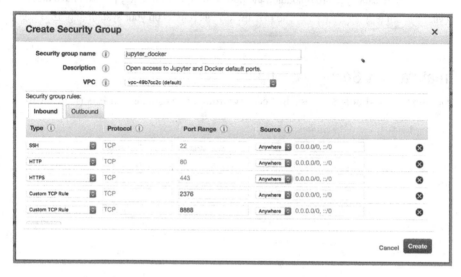

Figure 1-7. *Inbound rules for new security group*

Table 1-1. *Ports and Usages*

Port	Service Available
22	SSH
80	HTTP
443	HTTPS
2376	Docker Hub
8888	Jupyter

Create a New EC2 Instance

To create a new instance, start from the EC2 Dashboard and click the Launch Instance button (Figure 1-8).

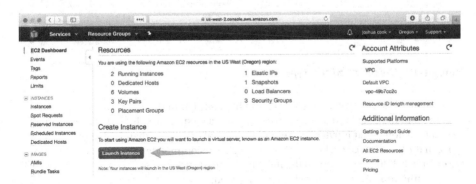

Figure 1-8. *Launch a new instance*

The launching of a new instance is a multi-step process that walks the user through all configurations necessary. The **first tab** is "Choose AMI." An AMI is an Amazon Machine Image[7] and contains the software you will need to run your sandbox machine. I recommend choosing the latest stable Ubuntu Server release that is free-tier eligible. At the time of writing, this was ami-efd0428f, Ubuntu Server 16.04 LTS (HVM), SSD Volume Type (Figure 1-9).

Figure 1-9. *Choose the latest stable Ubuntu Server release as AMI*

The **second tab** is "Choose Instance Type." In usage, I have found that the free tier, t2.micro (Figure 1-10), is sufficient for many applications, especially the sort of sandbox-type work that might be done in working through this text. This is to say that while working through the text, you may not be doing extended work on datasets, but rather learning about how to configure different systems. As such, your memory needs may be diminished. The ultimate goal is for the reader to be able to create and destroy machines at will. At this level of mastery, the reader can choose the minimum requirements for any application.

---

[7]http://docs.aws.amazon.com/AWSEC2/latest/UserGuide/AMIs.html

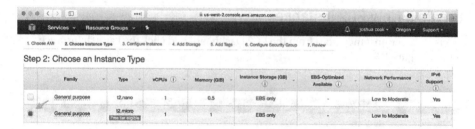

Figure 1-10. Use the `t2.micro` *type*

The **third tab,** "Configure Instance," can be safely ignored.

The **fourth tab** is "Add Storage." This option is also specific to intended usage. It should be noted that Jupyter Docker images can take up more than 5GB of disk space in the local image cache. For this reason, it is recommended to raise the value from the default 8GB to somewhere in the neighborhood of 20GB.

The **fifth tab,** "Add Tags," can be safely ignored.

The **sixth tab**, "Configure Security Group," is critical for the proper functioning of your systems. Previously, you configured a new security group to be used by your system. You will need to assign the security group that you just created, `jupyter_docker`, to the instance you are configuring. Choose "Select an **existing** security group," and then select the security group you just created. Verify that ports 22, 80, 443, 2376, and 8888 are available in the Inbound Rules at the bottom of the tab (Figure 1-11).

▨ **Note**   Most readers will receive a warning from AWS at this final phase that says something to the effect of "Improve your instances' security. Your security group, jupyter_docker, is open to the world." It is the opinion of this author that this warning can be safely ignored. The warning is letting us know that the instance we are preparing to launch can be accessed on the open web. This is intentional and by design. In this first and last conversation about system security, we will wave our hands at the concern and quickly move to the business of developing short-lifespan scalable systems.

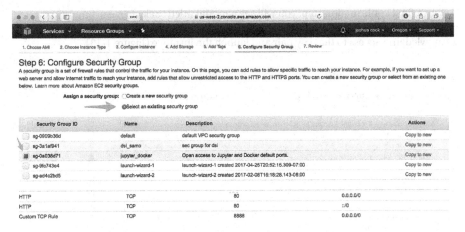

Figure 1-11. *Configure the security group for all traffic*

Finally, click "Review and Launch." Here, you see the specific configuration of the EC2 instance you will be creating. Verify that you are creating a t2.micro running the latest free tier-eligible version of Ubuntu Server and that it is available to all traffic, and then click the Launch button (Figure 1-12).

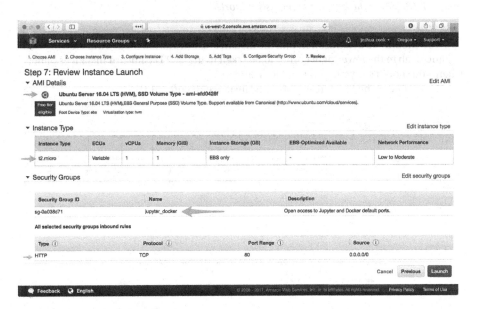

Figure 1-12. *Launch the new EC2 instance*

In a final confirmation step, you will see a modal titled "Select an existing key pair or create a new key pair." Select the key pair you previously created. Check the box acknowledging access to that key pair and launch the instance (Figure 1-13).

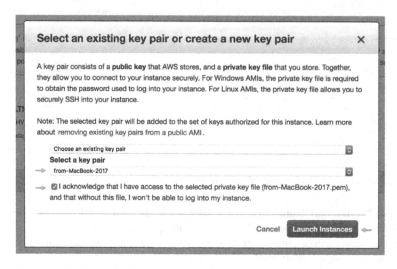

Figure 1-13. *Select the key pair previously imported*

You should see a notification that the instance is now running. Click the View Instances tab in the lower right corner to be taken to the EC2 Dashboard Instances pane, where you should see your new instance running.

Make note of the IP address of the new instance (Figure 1-14).

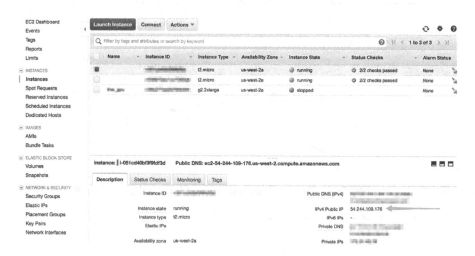

Figure 1-14. *Note the IP address of a running instance*

Configure the New EC2 Instance for Using Docker

Having set up an EC2 instance, you ssh into the instance using the IP you just obtained in order to provision the new instance with Docker (Listing 1-3).

Listing 1-3. SSH into EC2 Instance

```
$ ssh ubuntu@54.244.109.176
Welcome to Ubuntu 16.04.2 LTS (GNU/Linux 4.4.0-64-generic x86_64)
...
```

■ **Note** The first time you access your EC2 instance, you should see the following message: The authenticity of host '54.244.109.176 (54.244.109.176)' can't be established ... Are you sure you want to continue connecting (yes/no)? This is expected. You should hit <ENTER> to accept or type yes and hit <ENTER>.

Next (Listing 1-4), you install and configure Docker using a convenient install script provided by the Docker team. The script is obtained from get.docker.com and passed via pipe (|) to a shell (sh).

Listing 1-4. Install Docker Via a Shell Script

```
$ curl -sSL https://get.docker.com/ | sh
apparmor is enabled in the kernel and apparmor utils were already installed
+ sudo -E sh -c sleep 3; apt-get update
...
+ sudo -E sh -c docker version
Client:
 Version:      17.04.0-ce
 API version:  1.28
 Go version:   go1.7.5
 Git commit:   4845c56
 Built:        Mon Apr  3 18:07:42 2017
 OS/Arch:      linux/amd64

Server:
 Version:      17.04.0-ce
 API version:  1.28 (minimum version 1.12)
 Go version:   go1.7.5
 Git commit:   4845c56
 Built:        Mon Apr  3 18:07:42 2017
 OS/Arch:      linux/amd64
 Experimental: false

...
```

In Listing 1-5, you add the ubuntu user to the docker group. By default, the command line docker client will require sudo access in order to issue commands to the docker daemon. You can add the ubuntu user to the docker group in order to allow the ubuntu user to issue commands to docker without sudo.

Listing 1-5. Add the Ubuntu User to the Docker Group

```
$ sudo usermod -aG docker ubuntu
```

Finally, in order to force the changes to take effect, you reboot the system (Listing 1-6). As an alternative to rebooting the system, users can simply disconnect and reconnect to their remote system.

Listing 1-6. Restart the Docker Daemon

```
$ sudo reboot
```

The reboot will have the effect of closing the secure shell to your EC2 instance. After waiting a few moments, reconnect to the system. At this point, your system will be ready for use. sudo should no longer be required to issue commands to the docker client. You can verify this by connecting to your remote system and checking the dock version (Listing 1-7).

Listing 1-7. Log into the Remote System and Check the Docker Version

```
$ ssh ubuntu@54.244.109.176
$ docker -v
Docker version 17.04.0-ce, build 4845c56
```

Infrastructure Limitations on Data

Before commencing with the nuts and bolts of using Docker and Jupyter to build scalable systems for computational programming, let's conduct a simple series of experiments with this new AWS instance. You'll begin with a series of simple questions:

> *What size dataset is too large for a t2.micro to load into memory?*
> *What size dataset is so large that, on a t2.micro, it will prevent Jupyter from fitting different kinds of simple machine learning classification models[8] (e.g. a K Nearest Neighbor model)? A Decision Tree model? A Logistic Regression? A Support Vector Classifier?*

To answer these questions, you will proceed in the following fashion:

1. Run the jupyter/scipy-notebook image using Docker on your AWS instance.

2. Monitor memory usage at runtime and as you load each dataset using docker stats.

---

[8]http://scikit-learn.org/stable/tutorial/machine_learning_map/

3. Use the `sklearn.datasets.make_classification` function to create datasets of arbitrary sizes using a Jupyter Notebook and perform a fit.

4. Restart the Python kernel after each model is fit.

5. Take note of the dataset size that yields a memory exception.

Pull the `jupyter/scipy-notebook` image

Since you are working on a freshly provisioned AWS instance, you must begin by pulling the Docker image with which you wish to work, the `jupyter/scipy-notebook`. This can be done using the `docker pull` command, as shown in Listing 1-8. The image is pulled from Project Jupyter's public Docker Hub account.[9]

Listing 1-8. Pull the jupyter/scipy-notebook image.

```
ubuntu@ip-172-31-6-246:~$ docker pull jupyter/scipy-notebook
Using default tag: latest
latest: Pulling from jupyter/scipy-notebook
693502eb7dfb: Pull complete
a3782c2efb41: Pull complete
9cb32b776a40: Pull complete
e539f5722cd5: Pull complete
b4690d4047c6: Pull complete
121dc465f5c6: Pull complete
c352772bbcfd: Pull complete
eeda14d1c421: Pull complete
0057b9e76c8a: Pull complete
e63bd87d75dd: Pull complete
055904fbc069: Pull complete
d336770b8a83: Pull complete
d61dbef85c7d: Pull complete
c1559927bbf2: Pull complete
ee5b638d15a3: Pull complete
dc937a931aca: Pull complete
4327c0faf37c: Pull complete
b37332c24e8c: Pull complete
b230bdb41817: Pull complete
765fecb84d9c: Pull complete
97efa424ddfa: Pull complete
ccfb7ed42913: Pull complete
2fb2abb673ce: Pull complete
Digest: sha256:04ad7bdf5b9b7fe88c3d0f71b91fd5f71fb45277ff7729dbe7ae20160c7a56df
Status: Downloaded newer image for jupyter/scipy-notebook:latest
```

---

[9]http://hub.docker.com/u/jupyter/

Once you have pulled the image, it is now present in your docker images cache. Anytime you wish to run a new Jupyter container, Docker will load the container from the image in your cache.

Run the jupyter/scipy-notebook Image

In Listing 1-9, you run a Jupyter Notebook server using the minimum viable docker run command. Here, the -p flag serves to link port 8888 on the host machine, your EC2 instance, to the port 8888 on which the Jupyter Notebook server is running in the Docker container.

Listing 1-9. Run Jupyter Notebook Server

```
$ docker run -p 8888:8888 jupyter/scipy-notebook
[I 22:10:01.236 NotebookApp] Writing notebook server cookie secret to /home/
jovyan/.local/share/jupyter/runtime/notebook_cookie_secret
[W 22:10:01.326 NotebookApp] WARNING: The notebook server is listening on
all IP addresses and not using encryption. This is not recommended.
[I 22:10:01.351 NotebookApp] JupyterLab alpha preview extension loaded from
/opt/conda/lib/python3.5/site-packages/jupyterlab
[I 22:10:01.358 NotebookApp] Serving notebooks from local directory: /home/
jovyan/work
[I 22:10:01.358 NotebookApp] 0 active kernels
[I 22:10:01.358 NotebookApp] The Jupyter Notebook is running at: http://[all
ip addresses on your system]:8888/?token=7b02e3aadb29c42ff066a7290d81dd48e4
4ce62bd7f2bd0a
[I 22:10:01.359 NotebookApp] Use Control-C to stop this server and shut down
all kernels (twice to skip confirmation).
[C 22:10:01.359 NotebookApp]

    Copy/paste this URL into your browser when you connect for the first
    time, to login with a token: http://localhost:8888/?token=7b02e3aadb29c4
    2ff066a7290d81dd48e44ce62bd7f2bd0a.
```

The output from the running Jupyter Notebook server provides you with an authentication token (token=7b02e3aadb29c42ff066a7290d81dd48e44ce62bd7f2bd0a) you can use to access the Notebook server through a browser. You can do this using the URL provided with the exception that you will need to replace localhost with the IP address of your EC2 instance (Listing 1-10).

Listing 1-10. The URL of a Jupyter Instance Running on AWS with an Access Token Passed as a Query Parameter

```
http://54.244.109.176:8888/?token=1c32913725d84a76e7b3f04c45b91e17b77f
3c3574779101.
```

Monitor Memory Usage

In Listing 1-11, you have a look at your running container using the `docker ps` command. You will see a single container running with the `jupyter/scipy-notebook` image.

Listing 1-11. Monitor Running Docker Containers

```
$ docker ps
CNID  IMAGE        COMMAND    CREATED      STATUS     PORTS            NAMES
cfef  jupyter/
      scipy...     "tini..."  10 min ago   Up 10 min  0.0.0.0:8888->   friendly_
                                                      8888/tcp         curie
```

Next, you use `docker stats` to monitor the active memory usage of your running containers (Listing 1-12). `docker stats` is an active process you will use to watch memory usage throughout.

Listing 1-12. Monitor Docker Memory Usage.

```
$ docker stats
CONTAINER     CPU %  MEM USAGE / LIMIT   MEM %  NET I/O   BLOCK I/O    PIDS
cfef9714b1c5  0.00%  49.73MiB / 990.7MiB 5.02%  60.3kB /  10.4MB / 0B  2
                                                1.36MB
```

You can see several things here germane to the questions above. The server is currently using none of the allotted CPU.[10] You can see that the Docker container has nearly 1GB of memory available to it, and of this, it is using 5%, or about 50MB. The 1GB matches your expectation of the amount of memory available to a `t2.micro`.

What Size Data Set Will Cause a Memory Exception?

You are going to be using Jupyter Notebook to run the tests. First, you will create a new notebook using the Python 3 kernel (Figure 1-15).

Figure 1-15. *Create a new notebook*

---

[10]My `t2.micro` has but a single CPU.

17

In Listing 1-13, you examine your memory usage once more. (After launching a new notebook, the current memory usage increases to about 9% of the 1GB.)

Listing 1-13. Monitor Docker Memory Usage

```
$ docker stats
CONTAINER     CPU %   MEM USAGE / LIMIT    MEM %  NET I/O    BLOCK I/O   PIDS
cfef9714b1c5  0.01%   87.18MiB / 990.7MiB  8.80%  64.2kB /   12.6MB /    13
                                                  1.48MB     217kB
```

If you close and halt (Figure 1-16) your running notebook, you can see memory usage return to the baseline of about 5% of the 1GB (Listing 1-14).

Figure 1-16. *Close and halt a running notebook*

Listing 1-14. Monitor Docker Memory Usage

```
$ docker stats
CONTAINER     CPU %   MEM USAGE / LIMIT    MEM %  NET I/O    BLOCK I/O   PIDS
cfef9714b1c5  0.00%   55.31MiB / 990.7MiB  5.58%  109kB /    12.4MB /    4
                                                  1.56MB     397kB
```

The Python machine learning library scikit-learn[11] has a module dedicated to loading canonical datasets and generating synthetic datasets: sklearn.datasets. Relaunch your notebook and load the make_classification function from this module (Listing 1-15, Figure 1-17), using the standard Python syntax for importing a function from a module. Examine memory usage once more (Listing 1-16).

Listing 1-15. Import make_classification

```
In [1]: from sklearn.datasets import make_classification
```

---

[11]http://scikit-learn.org/

```
In [1]: from sklearn.datasets import make_classification
```

Figure 1-17. *Import* `make_classification`

Listing 1-16. Monitor Docker Memory Usage

```
$ docker stats
CONTAINER    CPU %  MEM USAGE / LIMIT    MEM %    NET I/O   BLOCK I/O   PIDS
cfef9714b1c5 0.04%  148.3MiB / 990.7MiB  14.97%   242kB /   49.3MB /    13
                                                  4.03MB    340kB
```

Next (Listing 1-17, Figure 1-18), you create a new classification dataset using the default values. You then use the %whos IPython magic command[12] to display the size of the dataset in memory. After this, you examine memory usage (Listing 1-18).

Listing 1-17. Create a New Classification Dataset Using Default Values

```
In [2]: X, y = make_classification()
In [3]: %whos
Variable            Type          Data/Info
--------------------------------------------------------
X                   ndarray       100x20: 2000 elems, type 'float64', 16000
                                                                       bytes
make_classification function      <function make_classification at 0x7feb19
                                                                     2669d8>
y                   ndarray       100: 100 elems, type 'int64', 800 bytes
```

```
In [1]: from sklearn.datasets import make_classification

In [2]: X, y = make_classification()

In [3]: %whos
        Variable            Type       Data/Info
        --------------------------------------------------------
        X                   ndarray    100x20: 2000 elems, type `float64`, 16000 bytes
        make_classification function   <function make_classification at 0x7f4e3165da60>
        y                   ndarray    100: 100 elems, type `int64`, 800 bytes
```

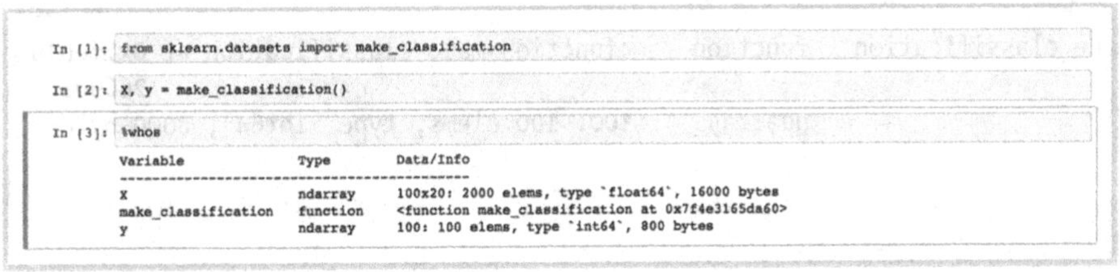

Figure 1-18. *Import* `make_classification`

Listing 1-18. Monitor Docker Memory Usage

```
$ docker stats
CONTAINER    CPU %  MEM USAGE / LIMIT    MEM %    NET I/O   BLOCK I/O  PIDS
cfef9714b1c5 0.01%  152MiB / 990.7MiB    15.35%   268kB /   54.1MB /   13
                                                  4.1MB     926kB
```

---

[12]https://ipython.org/ipython-doc/3/interactive/magics.html#magic-whos

19

So far you are minimally taxing your system. Take note of the size of the dataset, size in Python memory, and Docker system usage associated with this default classification dataset and then restart the Python kernel (Figure 1-19).

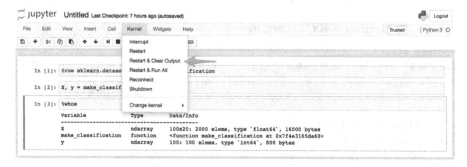

Figure 1-19. Restart the Python kernel

Next, you rerun the same experiment, increasing the size of your feature set by a factor of 10 (Listing 1-19, Figure 1-20). In Listing 1-20, you examine Docker system usage.

Listing 1-19. Create a New Classification Dataset

```
In [2]: X, y = make_classification(n_samples=1000, n_features=20)
In [3]: %whos
Variable                 Type         Data/Info
---------------------------------------------------
X                        ndarray      100x20: 2000 elems, type `float64`, 160000
                                                                            bytes
make_classification      function     <function make_classification at 0x7feb19
                                                                         2669d8>
y                        ndarray      100: 100 elems, type `int64`, 8000 bytes
```

```
In [1]: from sklearn.datasets import make_classification

In [2]: X, y = make_classification n_samples=1000, n_features=20)

In [3]: %whos
        Variable              Type        Data/Info
        -------------------------------------------------
        X                     ndarray     1000x20: 20000 elems, type `float64`, 160000 bytes (156.25 kb)
        make_classification   function    <function make_classification at 0x7fd5d3d6fae8>
        y                     ndarray     1000: 1000 elems, type `int64`, 8000 bytes
```

Figure 1-20. Import make_classification

Listing 1-20. Monitor Docker Memory Usage

```
$ docker stats
CONTAINER     CPU %   MEM USAGE / LIMIT     MEM %    NET I/O    BLOCK I/O    PIDS
cfef9714b1c5  0.01%   149.7MiB / 990.7MiB   15.11%   286kB /    54.5MB /     13
                                                     4.13MB     1.13MB
```

Repeat the experiment several more times, capturing the results in Table 1-2. Each time, restart the kernel, create a new dataset that is 10 times larger than the previous, and then examine the result in terms of memory usage using the IPython magic command %whos and the docker stats tool.

Table 1-2. *Classification Dataset Memory Footprint on t2.micro*

Shape of Feature Set	Size in Python Memory	Docker System Usage
100 × 20	.016MB	152MB
1000 × 20	.16MB	149.7MB
1000 × 200	1.525MB	152.8MB
10000 × 200	15.25MB	162.6MB
10000 × 2000	152.6MB	279.7MB
100000 × 2000	Memory Exception	N/A

Restart the Python kernel after each dataset is created. Take note of the dataset size that causes a memory exception.

When you attempt to create a classification dataset of size 100000 by 2000, you will hit a MemoryError, as seen in Listing 1-21 and Figure 1-21.

```
In [2]: X, y = make_classification(n_samples=100000, n_features=2000)
        ----------------------------------------------------------------------
        MemoryError                               Traceback (most recent call last)
        <ipython-input-2-df42c0ced9d5> in <module>()
        ----> 1 X, y = make_classification(n_samples=100000, n_features=2000)

        /opt/conda/lib/python3.5/site-packages/sklearn/datasets/samples_generator.py in make_classification(n_samples, n_feat
        ures, n_informative, n_redundant, n_repeated, n_classes, n_clusters_per_class, weights, flip_y, class_sep, hypercube,
        shift, scale, shuffle, random_state)
            179
            180     # Initialize X and y
        --> 181     X = np.zeros((n_samples, n_features))
            182     y = np.zeros(n_samples, dtype=np.int)
            183

        MemoryError:
```

Figure 1-21. *MemoryError when attempting to create a classification dataset*

Listing 1-21. MemoryError When Attempting to Create a Classification Dataset

```
In [2]: X, y = make_classification(n_samples=100000, n_features=2000)
----------------------------------------------------------------------
MemoryError                               Traceback (most recent call last)
<ipython-input-2-df42c0ced9d5> in <module>()
----> 1 X, y = make_classification(n_samples=100000, n_features=2000)

/opt/conda/lib/python3.5/site-packages/sklearn/datasets/samples_generator.
py in make_classification(n_samples, n_features, n_informative, n_redundant,
n_repeated, n_classes, n_clusters_per_class, weights, flip_y, class_sep,
hypercube, shift, scale, shuffle, random_state)
```

```
       179
       180      # Initialize X and y
  --> 181      X = np.zeros((n_samples, n_features))
       182      y = np.zeros(n_samples, dtype=np.int)
       183
```

MemoryError:

And with that, you have hit the memory ceiling for your current system. It is not a particularly large dataset: 100,000 rows and 2000 columns. But then again, you are not working with a particularly large system either: a single CPU and 1GB of RAM. Certainly, you can imagine situations in which you will want to work with larger datasets on larger systems.

What Size Dataset Is Too Large to Be Used to Fit Different Kinds of Simple Models?

Next, let's answer the second question. Let's do this by starting with a fresh Docker container. First, in Listing 1-22, you again use docker ps to display running containers.

Listing 1-22. Monitor Running Docker Containers

```
$ docker ps
CNID IMAGE      COMMAND     CREATED      STATUS      PORTS           NAMES
cfef jupyter/   "tini..."  53 min ago   Up 53 min   0.0.0.0:8888->  friendly_
     scipy...                                        8888/tcp        curie
```

In Listing 1-23, you stop and then remove this container.

Listing 1-23. Stop and Remove a Running Container

```
$ docker stop friendly_curie
friendly_curie
ubuntu@ip-172-31-1-64:~$ docker rm friendly_curie
friendly_curie
```

Next, in Listing 1-24, you launch a brand new jupyter/scipy-notebook container.

Listing 1-24. Run Jupyter Notebook Server

```
$ docker run -p 8888:8888 jupyter/scipy-notebook
[I 20:05:42.246 NotebookApp] Writing notebook server cookie secret to /home/
jovyan/.local/share/jupyter/runtime/notebook_cookie_secret
...

    Copy/paste this URL into your browser when you connect for the first time,
    to login with a token:
        http://localhost:8888/?token=7a65c3c7dc6ea294a38397a48cc1ffe110ea13
        8aef6d42c4
```

Make sure to take note of the new security token (7a65c3c7dc6ea294a38397a48 cc1ffe110ea138aef6d42c4) and again use the AWS instance's IP address in lieu of localhost (Listing 1-25).

Listing 1-25. The URL of the New Jupyter Instance Running on AWS with an Access Token Passed as a Query Parameter

```
http://54.244.109.176:8888/?token=7a65c3c7dc6ea294a38397a48cc1ffe110ea138ae
f6d42c4
```

Before you start, measure the baseline usage for this current container via docker stats (Listing 1-26).

Listing 1-26. Monitor Docker Memory Usage

```
$ docker stats
CONTAINER     CPU %  MEM USAGE / LIMIT     MEM %  NET I/O   BLOCK I/O   PIDS
22efba43b763  0.00%  43.29MiB / 990.7MiB   4.37%  768B /    0B / 0B     2
                                                  486B
```

You again create a new Python 3 Notebook and set out to answer this second question. In Listing 1-27, you examine the memory usage of your Docker machine with a brand new notebook running.

Listing 1-27. Monitor Docker Memory Usage

```
$ docker stats
CONTAINER     CPU %  MEM USAGE / LIMIT     MEM %  NET I/O   BLOCK I/O    PIDS
22efba43b763  0.04%  90.69MiB / 990.7MiB   9.15%  58.8kB /  0B / 217kB  13
                                                  1.3MB
```

The approach to solving this problem will be slightly different and will make heavier use of docker stats. The %whos IPython magic command cannot be used to display memory usage of a fit model and, in fact, a trivial method for measuring memory usage does not exist.[13] You will take advantage of your knowledge of the space in memory occupied by the data created by make_classification and this baseline performance you just measured.

You will use the code pattern in Listing 1-28 to perform this analysis.

Listing 1-28. Create a New Classification Dataset and Perform Naïve Model Fit

```
from sklearn.datasets import make_classification
from sklearn.<model_module> import <model>
X, y = make_classification(<shape>)
model = <model>()
model.fit(X, y)
model.score(X, y)
```

---

[13]https://stackoverflow.com/questions/449560/how-do-i-determine-the-size-of-an-object-in-python

23

For example, use the following code to fit the smallest KNeighborsClassifier (Listing 1-29), DecisionTreeClassifier (Listing 1-30), LogisticRegression (1-31), and SVC (Listing 1-32). You will then modify the scope of the data for each subsequent test.

Listing 1-29. Fit the smallest KNeighborsClassifier

```
from sklearn.datasets import make_classification
from sklearn.neighbors import KNeighborsClassifier
X, y = make_classification(1000, 20)
model = KNeighborsClassifier()
model.fit(X, y)
model.score(X, y)
```

Listing 1-30. Fit the smallest DecisionTreeClassifier

```
from sklearn.datasets import make_classification
from sklearn.tree import DecisionTreeClassifier
X, y = make_classification(1000, 20)
model = DecisionTreeClassifier()
model.fit(X, y)
model.score(X, y)
```

Listing 1-31. Fit the smallest LogisticRegression

```
from sklearn.datasets import make_classification
from sklearn.linear_model import LogisticRegression
X, y = make_classification(1000, 20)
model = LogisticRegression()
model.fit(X, y)
model.score(X, y)
```

Listing 1-32. Fit the smallest SVC

```
from sklearn.datasets import make_classification
from sklearn.neighbors import SVC
X, y = make_classification(1000, 20)
model = SVC()
model.fit(X, y)
model.score(X, y)
```

You then use docker stats to examine the Docker system usage. In between each test, you use docker restart (Listing 1-33) followed by the container id 22efba43b763 to reset the memory usage on the container. After restart the container, you will typically have to confirm restarting the Jupyter kernel as well (Figure 1-22).

Listing 1-33. Restart Your Docker Container

```
$ docker restart 22efba43b763
```

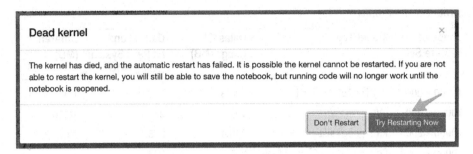

Figure 1-22. *Confirm a restart of the Jupyter kernel*

The results of this experiment are captured in Table 1-3 and Figure 1-23.

Table 1-3. *Classification of Dataset and Model Memory Footprint on* `t2.micro`

Shape of Feature Set	Model Type	Dataset System Usage (MB)	Dataset and Fit Peak System Usage (MB)	Difference (MB)
Baseline (No Notebook running)	N/A	N/A	N/A	40.98
Baseline (Notebook running)	N/A	N/A	N/A	76.14
100 × 20	KNeighborsClassifier	99.9	100.0	0.1
100 × 20	DecisionTreeClassifier	103.2	103.3	0.1
100 × 20	LogisticRegression	102.5	102.6	0.1
100 × 20	SVC	101.2	101.4	0.2
1000 × 20	KNeighborsClassifier	100.3	100.4	0.1
1000 × 20	DecisionTreeClassifier	103.7	103.8	0.1
1000 × 20	LogisticRegression	104.9	105.1	0.2
1000 × 20	SVC	104.9	105.5	0.6
1000 × 200	KNeighborsClassifier	106.3	106.9	0.6
1000 × 200	DecisionTreeClassifier	104.8	105.7	0.9
1000 × 200	LogisticRegression	102.0	102.3	0.3
1000 × 200	SVC	104.6	106.0	1.4
10000 × 200	KNeighborsClassifier	115.5	117.8	2.3
10000 × 200	DecisionTreeClassifier	119.8	127.8	8.0

(*continued*)

Table 1-3. (*continued*)

Shape of Feature Set	Model Type	Dataset System Usage (MB)	Dataset and Fit Peak System Usage (MB)	Difference (MB)
10000 × 200	LogisticRegression	121.3	122.6	1.3
10000 × 200	SVC	121.1	286.7	165.6
10000 × 2000	KNeighborsClassifier	256.4	275.1	18.7
10000 × 2000	DecisionTreeClassifier	257.1	333.6	76.5
10000 × 2000	LogisticRegression	258.6	564.9	306.3
10000 × 2000	SVC	256.3	491.9	235.6

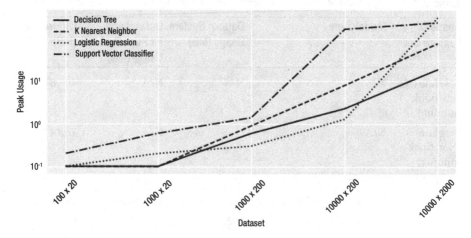

Figure 1-23. *Dataset vs. Peak Usage by model*

Measuring Scope of Data Capable of Fitting on T2.Micro

In the previous test, a 10000 row x 2000 column dataset was the largest that you were able to successfully load into memory. In this test, you were able to successfully fit a naïve implementation of four different machine learning models against each of the datasets that you were able to load into memory. That said, you can see that neither the LogisticRegression nor the SVC (Support Vector Classifier) are capable of handling much more.

Summary

In this chapter, I introduced the core subjects of this text, Docker and Jupyter, and discussed a recommended practice for working through this text, which is using a disposable Linux instance on Amazon Web Services. I provided detailed instructions for configuring, launching, and provisioning such an instance. Having launched an instance, you used Docker and Jupyter to explore a toy big data example, diagnosing memory performance as you loaded and fit models using a synthetic classification dataset generated by `scikit-learn`.

I did not intended for this chapter to have been the moment when you thoroughly grasped using Docker and Jupyter to build systems for performing data science. Rather, I hope that it has served as a substantive introduction to the topic. Rather than simply stating what Docker and Jupyter are, I wanted you to see what these two technologies are by *using* them.

In the chapters ahead, you will explore many aspects of the Docker and Jupyter ecosystems. Later, you will learn about the open source data stores Redis, MongoDB, and PostgreSQL, and how to integrate them into your Docker-based applications. Finally, you will learn about the Docker Compose tool and how to tie all of these pieces together in a single `docker-compose.yml` file.

Summary

CHAPTER 2

■ ■ ■

Docker

Docker is a way to isolate a process from the system on which it is running. It allows us to isolate the code written to define an application and the resources required to run that application from the hardware on which it runs. We add a layer of complexity to our software, but in doing so gain the advantage of ensuring that our local development environment will be identical to any possible environment into which we would deploy the application. If a system can run Docker, a system can run our process. With the addition of a thin layer of abstraction we become hardware independent. On its face, this would seem to be an impossible task. As of 2014, there were 285 actively maintained Linux distributions and multiple major versions of both OS X and Windows. How could we possibly write a system to allow for all possible development, testing, and production environments?

Docker solves this problem via **containerization**. We will never be able to guarantee that our remote environments will be running the same OS as we are locally. We often know for a fact that it never will (I develop using Mac OS X and usually deploy to systems running Ubuntu). That said, as visualized in Figure 2-1, we can guarantee that both our development and deployment environments will be able to run the Docker engine. We write our application to be run as a container by the Docker engine—we containerize our application—and thus ensure compatibility across platforms. We are not concerned about the underlying operating system or hardware, only that it is running the Docker engine.

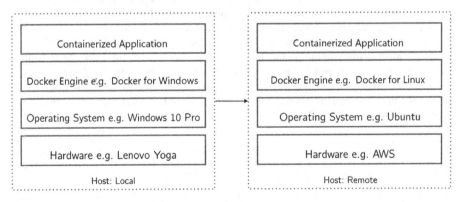

Figure 2-1. *Deploying across heterogenous infrastructure. Note that the technology stack might be completely different across systems, but we deploy an identical container.*

© Joshua Cook 2017

J. Cook, *Docker for Data Science*, DOI 10.1007/978-1-4842-3012-1_2

Using the suite of Docker tools we build our applications to run as a Docker container. Once built, we verify our images by running them on our local system. Having confirmed this, it is trivial to deploy our containerized application to a remote machine that is running the Docker engine. We run the system in exactly the same way regardless of operating system or hardware. Docker lets developers be OS agnostic.

Docker Is Not a Virtual Machine

Docker is not a Virtual Machine (VM) technology.[1] That said, it is useful to briefly look at what VM technology is. Many will be familiar with VM technology, especially as a VM being run using a tool such as VirtualBox. Using one of these tools, a single computer, the **host** runs many VM instances, **guests**. Each guest uses a large file on disk to define its isolated operating and file system. Each guest runs as a single, resource-intensive process on the host CPU.

For the purposes of most users, the guest behaves as a stand-alone computer, very similar in practical use to the host machine on which it runs. In other words, a VM feels like **virtualized hardware**. Downsides to VM technology include the consumption of large swaths of hard drive space to store a bulky operating system and the consumption of considerable CPU resources in maintaining all of the processes required of a full OS.

Containerization

Containerization is a virtualization method, but containers are not VMs in the way that most think of them. The confusion is understandable. Like containers, it is even possible to define VMs using the software Vagrant. But to be clear, a VM is using one kind of virtualization. Containerization is a different type of virtualization altogether.

The Linux Containers (LXC) project is a vendor-neutral project designed to provide a native set of tools and libraries for the containment of processes from the broader Linux system on which they are being run. LXC runs in the same operating system as its host. The stated goal of LXC is "to create an environment as close as possible to a standard Linux installation but without the need for a separate kernel." Put another way, LXC allows processes to be containerized.

Containerization seeks to virtualize processes. Thus, a containerized process is running in an environment optimized for its purposes, but is being run using the system resources of the host computer. The LXC library has been carefully designed to allow a containerized process to run as a *virtualized process* (see Figure 2-2) on its host system without the need to run a full operating system. LXCs have low overheads and better performance compared to VMs.

---

[1]https://blog.docker.com/2016/03/containers-are-not-vms/

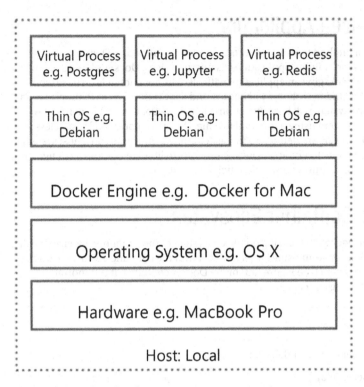

Figure 2-2. *A virtualized process*

In June of 2015, Docker helped to launch the Open Container Project. Docker donated its library runc to serve as an iteration upon LXC. While Docker is no longer running LXC at its core, the principle remains: Docker is not a virtual machine. Docker is leveraging virtualization technology to achieve the isolation of processes or services from the host systems.

Throughout this text, you will be consistently faced with two problems in running your processes via Docker: networking (connecting to your processes) and maintaining the persistence of data. With these latest innovations in managing containers, Docker is now moving toward stronger tools for managing both issues. With Docker managing networking and persistence, once these processes have been abstracted or containerized they can be run at will on any system.

A Containerized Application

On top of providing a method for running a containerized application, Docker also provides a set of tools for building applications as microservices. Docker's build system provides a system for packaging an application with all of its dependencies. Those who are comfortable with working with Ruby's bundler and Gemfile system or Python's conda and environment.yml system will be right at home in using a Dockerfile to define the requirements of their system using a minimal text file. Using this Dockerfile, stateless and immutable applications are defined to run as "compiled" processes on a host system running the Docker engine. In doing so, Docker attempts to liberate the software engineer from dependency on the hardware on which their code will run.

The Docker Container Ecosystem

You begin looking at Docker by focusing on the ecosystem of the container. Later, you will leverage Docker's tools for composing larger systems with the containers you have built. In the immediate ecosystem (see Figure 2-3) of the Docker container, it is important to keep track of the following concepts:

- The Docker CLI client

- The host

- The Docker engine or daemon

- The Docker image

- The Docker container

- The Docker registry, typically Docker Hub

- Docker Compose

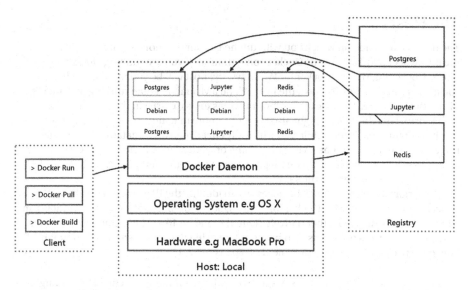

Figure 2-3. *The Docker ecosystem*

■ **Note**   You will eschew completely the use of the GUI tool Kitematic in favor of a wholly command line-oriented approach to managing Docker.

The Docker Client

The Docker client is a command line interface used to give instructions to the Docker engine regardless of the details of the engine's implementation on your system. This is similar to the client-server architecture of the Web, in which a client system uses an interface (typically a web browser, but possibly a RESTful API) to interact with a remote server. In the case of Docker, the Docker client talks to the Docker engine that performs the work of containers and containerization.

While you work through this text, you will be using the Docker command line interface as your client and the engine will be running on your local system or on a t2. micro as recommended in Chapter 1. Using the Docker client, you tell the Docker daemon to pull an image from a registry. You can then tell the Docker daemon to run that image. Having done so, you might ask the engine which ps or containers are currently running.[2]

To list commands available to the Docker client, either run docker with no parameters or execute docker help. Depending on your Docker system configuration, you may be required to preface each docker command with sudo. To avoid having to use sudo with the docker command, your system administrator can create a Unix group called docker and add users to it.

---

[2]This command is a descendent of the bash command ps. In bash, this command lists running processes, whereas in Docker this command lists running containers. This fits with our understanding of the Docker container as a virtualized process.

The Host

The host is a machine on which you will run the Docker daemon/engine. Locally, the host will depend upon your Docker configuration. If you are running Docker for Linux, Docker for Mac, or Docker for Windows, the host will be your system itself. If you are running Docker Toolbox, the host will be a virtual machine running on Oracle's VirtualBox software. The Docker Toolbox provides tools to assist in the proper creation of this virtual machine. You might also set up a remote machine to serve as your host. The important thing is that while you will always need a host, the details of this host are irrelevant. The host might be a virtual machine on your Mac, a c4.8xlarge EC2 instance on Amazon Web Services, or a bare metal server in your university basement. Regardless, your application will behave the same.

In certain situations, it may be necessary to identify the IP address of the host. This is typically not necessary with Docker for Linux, Docker for Mac, or Docker for Windows, in which Docker is running either natively (Linux) or native-like (Mac, Windows). In other cases, the IP address of the host can be identified using the docker-machine command line tool, specifically, docker-machine ls (see Listing 2-1).

Listing 2-1. Display Docker Hosts Associated with the Running Attached Docker Engine

```
$ docker-machine ls
NAME      ACTIVE    DRIVER       STATE      URL                        SWARM    DOCKER    ERRORS
default   -         virtualbox   Running    tcp://192.168.99.100:2376  v1.13.0
```

The Docker Engine

The Docker engine is a persistent process that manages containers. It is running as a background service or daemon on the host. In fact, the engine is occasionally referred to as the Docker daemon. The Docker engine does the core work of Docker: building, running, and distributing your Docker containers. In this text, you will interact with the engine directly but will do so through the Docker client. The power of Docker lies in your ability to work with the engine via the Docker client. You will hand the managing of your processes over to the Docker engine. If you can do so on one system, you can count on any work you do to behave the same when on any other machine that is capable of running the Docker engine.

The Docker Image and the Docker Container

Docker images are read-only. This is not to say that we can't make changes to an image, but that once we have made changes, what we have is a new kind of image. I like to think about languages with immutable data structures such as tuples in Python. Once you define a tuple, you can't modify it, although you can define a new tuple that takes the original and modifies it in some way.

34

The Docker engine has several methods for building our own images. These include the Docker client and via a domain-specific language (DSL) known as the `Dockerfile`. We can also download images that other people have created.

Docker containers are instances of Docker images. They are stand-alone, containing everything a single instance of an application needs to run (OS, dependencies, source, metadata), and can be run, started, stopped, moved, and deleted. They are also isolated and secure.

It is helpful to think of Docker images and containers in terms of object-oriented programming. An image is a defined "class" of container that we might create. A container is then an "instance" or "object" of that class. The Docker engine will manage multiple containers running on the host. When the engine runs a container from an image, it adds a read-write layer on top of the image in which our application can run.

Truthfully, however, this analogy of the image-container relationship as object-oriented programming is a weak analogy. A stronger analogy is the analogy of Docker as a compiled language. In this analog, we might think of a `Dockerfile` as source code, an image as a compiled binary, and a running container as a running process. It is clearly an intended analog considering that we use the command `docker ps` to display currently running containers. This analogy is helpful in a few ways. For one, it brings the build process to the forefront of our understanding of working with Docker. For another, it helps to reinforce the idea that containers are ephemeral, just like running processes. When they end their lifecycle, their state is effectively lost.

The Docker Registry

Docker registries hold images. These are public or private stores from which you upload or download images. For the purposes of this book, you will use the public Docker registry at Docker Hub.[3] Docker Hub is the source of the official images of the major open source technologies you will be using including Jupyter, PostgreSQL, Redis, and MongoDB.

Docker Compose

Docker Compose is a tool for assembling microservices and applications consisting of multiple containers. These microservices and applications are designed using a single YAML file and can be built, run, and scaled, each via a single command. Docker Compose is particularly useful for the data scientist in building standalone computational systems comprised of Jupyter and one or more data stores (e.g. Redis).

---

[3]`https://hub.docker.com`

Get Docker

As of the writing of this text there are four core ways to install Docker across the major operating systems:

- Docker for Linux

- Docker for Mac

- Docker for Windows

- Docker Toolbox

It is recommended that systems be configured to use at least 2GB of RAM. I have not encountered significant issues in allowing the Docker engine to use all available resources, like CPUs and RAM.

■ **Note**   I once more recommend that the reader use a sandbox system on a t2.micro as outlined in Chapter 1. Chapter 1 contains instructions for configuring an AWS instance running Ubuntu.

Docker for Linux

Docker for Linux runs natively on most major Linux operating systems. It is driven from the command line using bash or similar shell. It is useful to note that running Docker for Linux is the most common configuration used on remote servers. It is a worthwhile exercise for Mac and Windows developers to familiarize themselves with this configuration. That said, the experience is very similar across platforms. Detailed instructions for a list of available operating systems are available at https://docs.docker.com/engine/installation/. Here, I focus on installation instructions for an Ubuntu system. For installation on other Linux distributions, readers should refer to the link for specific instructions.

Installing Docker on an Ubuntu System

Installing Docker from the command line provides the highest degree of flexibility. On an Ubuntu system (here, I use Ubuntu 16.04.2), this can be done via the apt tool.

■ **Note**   These instructions are for installing Docker on Ubuntu. Users wishing to install Docker on another Linux variant should refer to the specific instructions for their system at https://docs.docker.com/engine/installation/.

In Listing 2-2, you use apt search to examine the packages associated with Docker that are available for installation via apt. You run apt update first to ensure that you have the latest list of available packages.

Listing 2-2. Display Docker Packages Available for Installation

```
$ apt update
$ apt search docker
Sorting... Done
Full Text Search... Done

...

docker/xenial 1.5-1 amd64
  System tray for KDE3/GNOME2 docklet applications

docker-compose/xenial 1.5.2-1 all
  Punctual, lightweight development environments using Docker

docker-doc/xenial-updates 1.12.6-0ubuntu1~16.04.1 all
  Linux container runtime -- documentation

docker-registry/xenial 2.3.0~ds1-1 amd64
  Docker toolset to pack, ship, store, and deliver content

docker.io/xenial-updates 1.12.6-0ubuntu1~16.04.1 amd64
  Linux container runtime

...
```

The package in which you have the most interest is the docker.io package. This package contains both the Docker daemon, also known as the Docker container runtime, and the Docker command line interface (CLI) executable. In Listing 2-3, you use apt policy to display meta-information available for the docker.io package. As of the writing of this text, the docker.io package available via apt is version 1.12.6.

Listing 2-3. Display Meta-Information for docker.io Package

```
$ apt policy docker.io
docker.io:
  Installed: (none)
  Candidate: 1.12.6-0ubuntu1~16.04.1
  Version table:
     1.12.6-0ubuntu1~16.04.1 500
        500 http://us-west-2.ec2.archive.ubuntu.com/ubuntu xenial-updates/
        universe amd64 Packages
     1.10.3-0ubuntu6 500
        500 http://us-west-2.ec2.archive.ubuntu.com/ubuntu xenial/universe
        amd64 Packages
```

■ **Note**   The 500 preceding each policy statement is a priority number and signifies that the package is "installable" on the system.

You begin your installation by ensuring that you have no older versions of docker installed (Listing 2-4). If you receive the message "Package 'docker' is not installed, so not removed," this means docker is not installed and you can proceed.

Listing 2-4. Remove Previous Installations of docker

```
$ sudo apt remove docker
Reading package lists... Done
Building dependency tree
Reading state information... Done
Package 'docker' is not installed, so not removed
0 upgraded, 0 newly installed, 0 to remove and 43 not upgraded.
```

Configure Docker Repository

To install Docker, you will use the Docker recommended best practice of installing from the Docker repository. In order to do this, you will first need to set up the repository. You will be doing so for Docker CE. First, in Listing 2-5, you will allow apt to use a repository over HTTPS.

Listing 2-5. Allow apt to Use a Repository Over HTTPS

```
$ sudo apt-get install \
>       apt-transport-https \
>       ca-certificates \
>       curl \
>       software-properties-common
Reading package lists... Done
Building dependency tree
Reading state information... Done
ca-certificates is already the newest version (20160104ubuntu1).
apt-transport-https is already the newest version (1.2.19).
curl is already the newest version (7.47.0-1ubuntu2.2).
software-properties-common is already the newest version (0.96.20.5).
0 upgraded, 0 newly installed, 0 to remove and 43 not upgraded.
```

Then, in Listing 2-6, you will add Docker's official GPG key.

Listing 2-6. Add Docker's Official GPG Key

```
$ curl -fsSL https://download.docker.com/linux/ubuntu/gpg | sudo apt-key add
-
OK
$ sudo apt-key fingerprint 0EBFCD88
pub    4096R/0EBFCD88 2017-02-22
       Key fingerprint = 9DC8 5822 9FC7 DD38 854A  E2D8 8D81 803C 0EBF CD88
uid                    Docker Release (CE deb) <docker@docker.com>
sub    4096R/F273FCD8 2017-02-22
```

Finally, in Listing 2-7, you add the appropriate Docker repository for your system architecture.

Listing 2-7. Add Your System's Specific Docker Repository

```
$ sudo add-apt-repository \
>    "deb [arch=$(dpkg --print-architecture)] https://download.docker.com/
linux/ubuntu \
>    $(lsb_release -cs) \
>    stable"
```

Install from Docker Repository

Having configured the Docker repository, you can install using the canonical apt update (Listing 2-8) and apt install (Listing 2-9).

Listing 2-8. Update the apt Registry

```
$ sudo apt update
...
Get:6 https://download.docker.com/linux/ubuntu xenial/stable amd64 Packages
[1,479 B]
...
43 packages can be upgraded. Run 'apt list --upgradable' to see them.
```

Listing 2-9. Install docker.

```
$ sudo apt install -y docker-ce
Reading package lists... Done
...
Preparing to unpack .../docker-ce_17.03.1~ce-0~ubuntu-xenial_amd64.deb ...
Unpacking docker-ce (17.03.1~ce-0~ubuntu-xenial) ...
...
Processing triggers for ureadahead (0.100.0-19) ...
```

Manage Docker as a Non-Root User

On Linux systems, the Docker engine binds to a Unix port rather than a TCP port. This port is typically owned by root and must be accessed via sudo in order to receive commands from the Docker client. A common pattern for Docker users on Linux systems is to create a docker group and add users to the group, as demonstrated in Listing 2-10. When the Docker daemon is restarted, it makes the port bound to the Docker engine read/writeable by the docker group. The outcome is that users are able to issue commands to the engine without prepending sudo to their commands. Note that this is not, strictly speaking, necessary.

Listing 2-10. Create a docker Group on Linux Systems

```
$ sudo groupadd docker $ sudo usermod -aG docker $USER
```

Log out and back in to see the changes take effect. At this point, users will be able to issues commands to the engine via docker without issuing sudo.

Docker for Mac

Docker for Mac runs Docker using the HyperKit[4] VM. It is driven from the command line using bash or similar shell. Detailed instructions as well as specific system requirements are available at https://docs.docker.com/docker-for-mac/. The stable channel for installation is recommended.

Scripts for bash completion come prepackaged with the Docker for Mac application. To activate bash completion, simply symlink these files to your bash_completion.d/ directory (see Listing 2-11).

Listing 2-11. Symlink Bash Completion Files on Mac OS X

```
$ ln -s /Applications/Docker.app/Contents/Resources/etc/docker.bash-
completion /usr/local/etc/bash_completion.d/docker
$ ln -s /Applications/Docker.app/Contents/Resources/etc/docker-machine.bash-
completion /usr/local/etc/bash_completion.d/docker-machine
$ ln -s /Applications/Docker.app/Contents/Resources/etc/docker-compose.bash-
completion /usr/local/etc/bash_completion.d/docker-compose
```

Docker for Windows

Docker for Windows runs using Microsoft Hyper-V. It is driven from the command line using PowerShell. Detailed instructions as well as specific system requirements are available at https://docs.docker.com/docker-for-windows/. Installing Docker for Windows can be somewhat challenging. It should be noted that Docker for Windows can only be used on Windows 10 Pro or Enterprise 64-bit operating systems and requires

- 64-bit processor with Second Level Address Translation (SLAT)
- 4GB system RAM at minimum
- BIOS-level hardware virtualization support

The following notes have been helpful during past installations:

- Make sure to press OK if prompted to enable Hyper-V during the Docker install.
- Select "Shared Drives" from the Docker settings and make sure to share the C: drive. This can be done via Docker settings (Figure 2-4).

---

[4]https://github.com/docker/HyperKit/

Figure 2-4. *Access Docker settings*

- If necessary, disable the firewall or create an exception.

- Make sure to use Windows PowerShell to access the issue commands to the engine.

Docker Toolbox

Docker Toolbox is available for older Mac or Windows systems that do not meet the requirements of the more natively implemented Docker for Mac or Docker for Windows. Docker Toolbox includes several tools:

- Docker Machine for running docker-machine commands

- Docker Engine for running the docker commands

- Docker Compose for running the docker-compose commands

- Kitematic, the Docker GUI

- A shell preconfigured for a Docker command-line environment

- Oracle VirtualBox

Installation instructions are available here: https://docs.docker.com/toolbox/ overview/. Docker Toolbox users will need to use the Docker Quickstart Terminal command line environment to issue commands to the Docker engine. Installation of Docker Toolbox will create a local docker-machine using VirtualBox that serves as the host.

Hello, Docker!

Minimally, using Docker to run your code consists of the following:

1. Pull a precompiled image or build an image from a Dockerfile.

2. Run the image as a new container.

If you have just installed Docker for this first time, you might try some minimal commands as verification that the Docker client is correctly installed and available on your path. Listings 2-12, 2-13, and 2-14 demonstrate three ways that this can be done: docker version, docker help, or which docker work well as a minimal test.

Listing 2-12. Display the Docker Version

```
$ docker version
sudo docker version
Client:
 Version:      17.03.1-ce
 API version:  1.27
 Go version:   go1.7.5
 Git commit:   c6d412e
 Built:        Mon Mar 27 17:14:09 2017
 OS/Arch:      linux/amd64

Server:
 Version:      17.03.1-ce
 API version:  1.27 (minimum version 1.12)
 Go version:   go1.7.5
 Git commit:   c6d412e
 Built:        Mon Mar 27 17:14:09 2017
 OS/Arch:      linux/amd64
 Experimental: false
```

■ **Note** Running this command gives information on the version of docker running on both the Docker client and the server.

Listing 2-13. Display Docker Help

```
$ docker help

Usage: docker COMMAND

A self-sufficient runtime for containers

...
```

Listing 2-14. Display the Location of the Docker Binary.

```
$ which docker
/usr/local/bin/docker
```

Having verified that the Docker client is properly installed, you can move on to the canonical "Hello, World!" as demonstrated in Listing 2-15.

Listing 2-15. Run the hello-world Image

```
$ docker run hello-world
Unable to find image 'hello-world:latest' locally
latest: Pulling from library/hello-world
78445dd45222: Pull complete
Digest: sha256:c5515758d4c5e1e838e9cd307f6c6a0d620b5e07e6f927b07d05f6d12a1ac8d7
Status: Downloaded newer image for hello-world:latest

Hello from Docker!
```

This message shows that your installation appears to be working correctly. To generate this message, Docker took the following steps:

1. The Docker client contacted the Docker daemon.

2. The Docker daemon pulled the hello-world image from the Docker Hub.

3. The Docker daemon created a new container from that image, which runs the executable that produces the output you are currently reading.

4. The Docker daemon streamed that output to the Docker client, which sent it to your terminal.

To try something more ambitious, you can run an Ubuntu container with

```
$ docker run -it ubuntu bash
```

Share images, automate workflows, and more with a free Docker ID from https:// cloud.docker.com/.

For more examples and ideas, visit https://docs.docker.com/engine/userguide/.

And with that, you have verified that Docker is correctly installed and functioning. When you execute this command, the Docker client sends the run hello-world command to the Docker engine. The Docker engine then does the following:

1. Checks for the hello-world image in your local cache of images.

2. If the image does not exist locally, downloads the image from Docker Hub.

3. Creates a new container using the image.

4. Allocates a filesystem and adds a read-write layer to the top of the image.

5. Sets up an IP address for the system.

6. Executes the shell command /hello as specified in the image's Dockerfile.

7. Upon completion of this process, terminates the container and shuts down.

Listing 2-16 demonstrates sort of an elemental Docker command: run the latest Ubuntu image, (run ubuntu) and connect to it via a bash shell (-i -t /bin/bash). When you execute this command, the Docker client sends the command to the Docker engine. The Docker engine does the following:

1. Checks for the ubuntu image in your local cache of images.

2. Downloads the image from Docker Hub, unless the image exists locally.

3. Creates a new container using the image.

4. Allocates a filesystem and adds a read-write layer to the top of the image.

5. Sets up an IP address for the system.

6. Executes the process /bin/bash within the container.

7. Connects you via your current terminal to the running /bin/ bash process.

Listing 2-16. Run the Base ubuntu Image and Connect to It Via Shell

```
$ docker run -it ubuntu  /bin/bash
root@eb5f4278d040:/#
```

You will be connected to the running ubuntu container until you shut down. You can interact (see Listing 2-17) with this process as though it were your native Ubuntu system to which you were connected via a bash shell.

Listing 2-17. Interact with Ubuntu Running as a Docker Container.

```
root@8b9461e1d7dd:/# ls
bin   dev  home  lib64  mnt  proc  run   srv  tmp  var
boot  etc  lib   media  opt  root  sbin  sys  usr
root@8b9461e1d7dd:/# ps
  PID TTY          TIME CMD
    1 ?        00:00:00 bash
   12 ?        00:00:00 ps
root@8b9461e1d7dd:/# ps aux
```

```
USER        PID %CPU %MEM   VSZ   RSS TTY      STAT START   TIME COMMAND
root          1  0.0  0.1 18240  3212 ?        S<s  04:00   0:00 /bin/bash
root         13  0.0  0.1 34424  2808 ?        R<+  04:01   0:00 ps aux
root@8b9461e1d7dd:/# exit
joshuacook@LOCALHOST:~$
```

Here, you ended our session by typing Ctrl+D as if you were connected to a remote system via ssh. In doing so, you have control returned to your local system, the host on which Docker is running.

It is useful to take note of the state while your Ubuntu image was running. It is not unusual that ls would show a complete standard Linux filesystem. It is not unusual that ps would return just a few items. It is highly unusual that ps aux would return two items. ps aux shows (a) processes for all users, (u) showing the owner of the process, and (x) including processes that are not attached to any terminal. In other words, in running ps aux, you have effectively shown all of the processes currently running on the system. Again, it is highly unusual that only processes running on the system are the shell through which you have connected (PID 1 /bin/bash) and the ps aux you are using to display running processes (PID 13 ps aux). Let that sink in. Essentially, your Docker container is running one process. More on this later.

Basic Networking in Docker

The final introductory piece you will examine before proceeding is networking in Docker. Many of the containers you will be working with will need to be accessed from the host using a network protocol such as TCP or HTTP. Luckily, Docker handles networking for us. In a minimal sense, you will manage networking via Docker by publishing ports.

In publishing a port, you explicitly bind a port or range of ports from a running container to the host. This is done via the (lowercase p) –p flag. This command makes explicit a connection between the host and the container. As such, it can only be defined as an argument passed to the Docker engine.

The pattern used in publishing a port is -p host_port:container_port. Let's say, for example, that I run a Flask[5] app, defined in a container called my_flask_app, on port 5000, via the command docker run -p 7777:5000 my_flask_app. In this case, I am publishing the port 5000 in the container on the port 7777 on the host. In other words, whatever is available on the container at port 5000 will be available on the host at port 7777. The effect of this to me as the local end user is that I can access the Flask app I have defined in my browser at http://localhost:7777 (or port 7777 on the host's IP if I am using Docker Toolbox).

The Python module SimpleHTTPServer can be used to run a file server to the directory from which it is launched. In Listing 2-18, you use a single command to launch a file server via a Docker container. The Docker daemon pulls the python image and uses it to run the Python module using the command python -m SimpleHTTPServer. The server runs on the default port of 8000 within the container. You link this to the port 5000 on your host and are able to access the file server through the browser (Figure 2-5) at http://localhost:8000.

---

[5]http://flask.pocoo.org

Listing 2-18. Launch the File Server Via the Docker Container

```
$ docker run -v ~:/home -p 5000:8000 python:2.7 python -m SimpleHTTPServer
Unable to find image 'python:2.7' locally
2.7: Pulling from library/python
6d827a3ef358: Already exists
2726297beaf1: Pull complete
7d27bd3d7fec: Pull complete
44ae682c18a3: Pull complete
824bd01a76a3: Pull complete
69702776c399: Pull complete
7be4e7612dd4: Pull complete
Digest: sha256:bda277274d53484e4026f96379205760a424061933f91816a6d66784c5e8
afdf
Status: Downloaded newer image for python:2.7
172.17.0.1 - - [16/Apr/2017 14:58:54] "GET / HTTP/1.1" 200 -
```

■ **Note**   There are subtle nuances to running Docker on disparate systems.
In most cases, Docker Toolbox users will be able to access the simple file server at
http://192.168.99.100:5000. Docker for Linux/Mac/Windows users should use
http://localhost:5000.

Directory listing for /

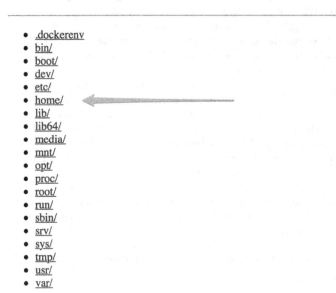

- .dockerenv
- bin/
- boot/
- dev/
- etc/
- home/
- lib/
- lib64/
- media/
- mnt/
- opt/
- proc/
- root/
- run/
- sbin/
- srv/
- sys/
- tmp/
- usr/
- var/

Figure 2-5. *Local file servers available via browser*

Summary

In this chapter, I formally introduced Docker and its ecosystem. I defined containerization and how it is useful to our work. I provided instructions for installing Docker on a few popular operating systems. Finally, you ran the Docker hello-world image and the Docker python image, using the latter to run a simple web server.

Having completed this chapter, I hope that you have an understanding of what exactly Docker is and some understanding of why it exists. I hope that you are aware of the various components of the Docker ecosystem. Readers should be able to run basic commands via the Docker client such as docker help, docker ps, and docker run.

In the next chapter, I will formally introduce Jupyter. As with everything in this text, I will be doing so using Docker. In subsequent chapters, you will explore in greater depth individual components of the Docker ecosystem, such as the Docker engine (or daemon), the Dockerfile, and Docker Hub.

CHAPTER 3

■ ■ ■

Interactive Programming

Interactive computing is a dialog between people and machines.

—Beki Grinter[1]

Jupyter is a web-based interactive application. Jupyter is a presentation environment. Jupyter is a new paradigm in programming. Jupyter is a way to save complex terminal sessions. All of this is to say that Jupyter is many things and, in this author's humble opinion, one of the most exciting innovations in computing in recent years. Jupyter is fundamentally changing the way we write code. To be sure, Jupyter doesn't replace vim, Sublime Text, or PyCharm. Jupyter replaces if __name__ == "__main__":.

Jupyter as Persistent Interactive Computing

We might see Jupyter as one part of a larger trend in the way we all engage with technology. We all actively interact with our computers every day. We pass a query to a search engine or to an online mapping tool and expect an immediate response. We change data in the cells and columns of a spreadsheet program and expect connected cells and columns to update instantaneously. We ask a voice interface in our connected home to dim the lights or play the latest pop song and expect immediate results. This is what we expect of our computers, and in this expectation it is easy to lose sight of the fact that this is fundamentally different from classical computer science.

How Not to Program Interactively

In order to illustrate the difference, it is useful to have a look at the C programming language. C is a descendent of Fortran and an ancestor of Python, all three of which live on today in practical everyday use. Many computational engineers do their work using Python and its computational libraries numpy and scipy, blissfully unaware of the fact that their core language is actually a high-level wrapper to C and their numerical libraries are actually high-level wrappers to the Fortran libraries BLAS and LAPACK. Never mind that they are totally unaware of what a development workflow looked like using C and BLAS. Until the somewhat recent advent of numpy, the computational engineer was forced to use a very different workflow.

---

[1] https://beki70.wordpress.com/2011/01/27/what-is-interactive-computing/

© Joshua Cook 2017
J. Cook, *Docker for Data Science*, DOI 10.1007/978-1-4842-3012-1_3

C is a compiled language. A functioning program, even one performing a simple mathematical calculation, requires considerable steps to bring it to life. Source code files must be written. This source code often includes references to header files existing elsewhere on the system. The code must be compiled into a binary including references to these header files and linked to any underlying compiled libraries used by the binary. Finally, the binary is executed, returning the results of the computation.

We will briefly demonstrate how this process works on a trivial calculation using the GNU Scientific Library (GSL). The GSL is a software library written in C used for computational mathematics. The GSL sits on top of a lower-level implementation of BLAS providing a layer of abstraction for ease in the development process.

Setting Up a Minimal Computational Project

As you are going to have more than one type of file to keep track of, it would behoove you to have a well-defined project directory. You first define a folder to hold your work, ch_3_ minimal_comp, and three subdirectories, bin, docker, and src, to your compiled binaries, Dockerfile, and source code, respectively. Upon completing this, on a Unix-like system, you might use the command tree to display your project hierarchy. If you do not have tree installed, you can do so using brew, apt, yum, or another package manager (see Listing 3-1), depending upon your system.

Listing 3-1. Install the Tree

```
$ # change apt to brew, yum, or the appropriate package manager for your
system
$ apt install tree
```

Next, in Listing 3-2, you make the directories necessary to begin your project. This is followed by Listing 3-3, in which you use the tree tool to display the project repository you have created.

Listing 3-2. Make the Project Directories

```
$ mkdir ch_3_minimal_comp ch_3_minimal_comp/bin ch_3_minimal_comp/src ch_3_
minimal_comp/docker
```

Listing 3-3. Display the Current State of Your Minimal Computational Project

```
$ tree
.
└── ch_3_minimal_comp
    ├── bin
    ├── docker
    └── src
```

Writing the Source Code for the Evaluation of a Bessel Function

Next, you create a new file, bessel.c, to be stored in the src directory, containing your source code (see Listing 3-4). Note that it includes two include statements, references to external code files containing extended functionality you will use in this program, stdio.h and gsl/gsl_sf_bessel.h. You will need to explicitly tell your compiler how to handle these files using an include statement and a linking statement.

Listing 3-4. Evaluation of the Zero-Order Bessel Function of the First Kind at $x=5$

```
// src/bessel.c
#include <stdio.h>
#include <gsl/gsl_sf_bessel.h>

int main(void)
{
  double x = 5.0;
  double y = gsl_sf_bessel_J0(x);
  printf("J0(%g) = %.18e\n", x, y);
  return 0;
}
```

Performing Your Calculation Using Docker

The completion of this calculation will require two separate processes. You will need to compile your source code into a binary and then execute this binary. As before, you will let Docker manage these processes. You will do so using a docker image defined by the Dockerfile shown in Listing 3-5.

■ **Note**   This image is built as an addition to the gcc image available on Docker Hub at https://hub.docker.com/_/gcc/.

Listing 3-5. GSL Dockerfile

```
FROM gcc

LABEL maintainer=@joshuacook

RUN apt-get update && \
    apt-get install -y \
    gsl-bin \
    libgsl0-dbg \
    libgsl0-dev \
    libgsl0ldbl
```

51

You save this file as Dockerfile in your docker directory. After doing this, you again use the tree command to display the state of your project (Listing 3-6).

Listing 3-6. Display the Current State of Your Minimal Computational Project

```
$ cd ch_3_minimal_comp
$ tree
.
├── bin
├── docker
│   └── Dockerfile
└── src
    └── bessel.c
```

You build the image for use with the docker build command. In your shell, you should currently be in the ch_3_minimal_comp directory. This is critical because your docker build command will make relative reference to a Dockerfile containing the definition of the image you wish to build. Here, you give the image a name via the -t flag, gsl. You specify that the build process should look for a Dockerfile within the relatively referenced directory docker. Listing 3-7 shows the complete command executed.

Listing 3-7. Build the gsl Docker Image

```
$ docker build -t gsl docker
Sending build context to Docker daemon 2.048 kB
Step 1/3 : FROM gcc
latest: Pulling from library/gcc
693502eb7dfb: Already exists

...

Status: Downloaded newer image for gcc:latest
 ---> 408d218617ca
Step 2/3 : MAINTAINER @joshuacook
 ---> Running in 43a89bcd4fae
 ---> 97faa5ea6f1e
Removing intermediate container 43a89bcd4fae
Step 3/3 : RUN apt-get update &&     apt-get install -y     gsl-bin
libgsl0-dbg     libgsl0-dev     libgsl0ldbl
 ---> Running in 988614cb6d56

...
 ---> 568814736d4b
Removing intermediate container 988614cb6d56
Successfully built 568814736d4b
```

When the build completes, this image will be available for use globally. You can verify this via the docker images command (see Listing 3-8).

Listing 3-8. Display Local Docker Images

```
$ docker images
REPOSITORY      TAG         IMAGE ID        CREATED         SIZE
gsl             latest      b3f3b5f49e4a    24 hours ago    1.52 GB
...
```

Compile Your Source Code

Next, you compile your source code. To compile, you will use the following docker run outlined in Listing 3-9.

Listing 3-9. Compile the Bessel Function Binary

```
$ docker run \
  -v 'pwd':/home \
  gsl \
  gcc \
  -I /usr/include/ \
  -L /usr/lib/ -lgsl -lgslcblas \
  /home/src/bessel.c
  -o /home/bin/bessel
```

This command instructs the Docker engine to

- Run a container.

- Attach the current working directory ('pwd') to the location /home within the container.

- Use the gsl image as a basis.

■ **Note** You make use of bash's inline execution functionality via two backticks. Executing pwd prints your working directory (your current location). Running this in backticks substitutes your working directory in place. Thus, -v 'pwd':/home attaches your working directory to the directory /home in your docker container.

Users interacting with the Docker engine using PowerShell, that is, Docker for Windows users, will need to use the alternative ${pwd} (e.g. -v ${pwd}:/home).

References to file locations are made relative to where they will be mounted within the container. The container will

- Run a single process, the Gnu C compiler, gcc.

- Include a reference to the location of the gsl header files (-I /usr/include/).

- Link the gsl libraries (-L /usr/lib/ -lgsl -lgslcblas).

- Use /home/src/bessel.c as source.

- Output (-o) a binary to /home/bin/Bessel.

The process should complete very quickly. When the Docker engine attached your directory as a volume, a two-way connection was created. Because of this, the compiled binary is now present in your host machine's file system. In Listing 3-10, you once more display the state of your project using the tree tool.

Listing 3-10. Display the Current State of Your Minimal Computational Project

```
$ tree
.
├── bin
│   └── bessel
├── docker
│   └── Dockerfile
└── src
    └── bessel.c
```

Execute Compiled Binary

Finally, you execute your compiled binary to retrieve the value of your calculation. Again, you must ask the Docker engine to manage this process for you, if for no other reason than that the binary was compiled to run on a gsl container.

It is worth emphasizing this last point. You could be working through these exercises on literally any operating system and hardware configuration. You are working with Docker to abstract your work away from your specific configuration. This means, however, that your binary has not been compiled to run on your local hardware/software configuration. It has been compiled to run on a gsl Docker container defined by the gsl Docker image. In this case, it is compiled to run on debian:jesse.

In Listing 3-11, you execute your compiled binary using Docker.

Listing 3-11. Execute the Bessel Function Binary

```
$ docker run -v 'pwd':/home gsl_image /home/bin/bessel
J0(5) = -1.775967713143382642e-01
```

But what if you wish to know the value at $x=6$ or $x=7$? In this case, you would need to go through the entire process once more: edit code, compile code, execute code. While there are significant advantages to writing code to run using the GSL, rapid iteration during the development cycle is not one of them and interaction is an impossibility. This entire process has been a demonstration of **how not to code interactively**.

How to Program Interactively

Jump 30 years forward. Put that compiler down! Thanks to the work of pioneers like Travis Oliphant and Fernando Perez, on numpy and iPython, we can perform this trivial calculation in the way that it was meant to be performed: trivially. numpy (I say "num"-"pie" to my elders and "num"-"pee" to my peers) is short for numerical Python and provides us with a high-level wrapper to a lower-level implementation of BLAS. IPython is short for interactive Python and is a highly-evolved Python REPL (read-eval-print loop) with a set of tools for interacting with any and all Python libraries.

■ **Note**   The reader should be careful not to confuse IPython, the command line REPL, and IPython Notebook, the legacy notebook server that has evolved into Jupyter.

Launch IPython Using Docker

Let's perform this trivial calculation. In Listing 3-12, you activate ipython using the jupyter/scipy-notebook Docker image.

Listing 3-12. Launch ipython Using the jupyter/scipy-notebook Image

```
$ docker run -it jupyter/scipy-notebook ipython
Python 3.5.2 |Continuum Analytics, Inc.| (default, Jul  2 2016, 17:53:06)
Type "copyright", "credits" or "license" for more information.

IPython 5.1.0 -- An enhanced Interactive Python.
?         -> Introduction and overview of IPython's features.
%quickref -> Quick reference.
help      -> Python's own help system.
object?   -> Details about 'object', use 'object??' for extra details.
```

In Listing 3-13, you perform the same calculation you performed earlier, although your process is not so elaborate. You first import the libraries necessary to evaluate a Bessel function in numpy, much as you imported the header files stdio.h and gsl/gsl_sf_bessel.h in your bessel.c file. Here, you import the special module from the scipy library. The special module contains a function called jv that is used to evaluate a "Bessel function of the first kind of real order and complex argument." You pass the arguments (0,5), signifying that you wish to evaluate the zero-order function at 5.

Listing 3-13. Evaluation of the Zero-Order Bessel Function of the First Kind at $x=5$

```
In [1]: import scipy.special as spc

In [2]: spc.jv(0, 5)
Out[2]: -0.17759677131433838
```

Performing these computations in an interactive REPL, your IPython environment, it is trivial to run the function twice more. **This** is interactive programming! In Listing 3-14, you run the functions with the arguments $(0,6)$ and $(0,7)$.

Listing 3-14. Evaluation of the Zero-Order Bessel Function of the First Kind at $x=6$ and $x=7$

```
In [3]: spc.jv(0, 6)
Out[3]: 0.15064525725099703

In [4]: spc.jv(0, 7)
Out[4]: 0.30007927051955563
```

Having perfomed all of the computations you set out to perform, you press Ctrl+D twice to exit. In doing so, you terminate the ipython process (being managed by your Docker daemon) and the container shuts down.

Persistence

With that your process has died and your container has shut down. For your intents and purposes your work is lost forever. There is real power to interaction but not if we have to repeat everything we have done every time we step away and return. We could write a short Python module that performs these calculations and then use Python (via Docker) to execute the module. In Listing 3-15, you write a short file named bessel.py that performs the same three computations you just performed interactively.

Listing 3-15. Three Bessel Function Calculations

```
import scipy.special as spc
print(spc.jv(0, 5))
print(spc.jv(0, 6))
print(spc.jv(0, 7))
```

In Listing 3-16, you once more display the state of your project using the tree tool.

Listing 3-16. Display the Current State of Your Minimal Computational Project

```
$ tree
.
├── bin
│   └── bessel
├── docker
│   └── Dockerfile
└── src
    ├── bessel.c
    └── bessel.py
```

In Listing 3-17, you execute your file, bessel.py.

■ **Note**   In order to run this, you must mount the directory containing your files as a volume to the location where the `jupyter/scipy-notebook` will look for files, in this case `/home/jovyan`.

Listing 3-17. Execute bessel.py Using Docker

```
$ docker run -v 'pwd'/src:/home/jovyan/work jupyter/scipy-notebook python
bessel.py
-0.177596771314
0.150645257251
0.30007927052
```

But now you are back to not programming interactively.

Jupyter Notebooks

Jupyter Notebooks are the evolution of IPython. Jupyter allows users to combine live code, markdown- and latex-rich text, images, plots, and more in a single document. As the successor to the IPython notebook, Jupyter was renamed as the platform began to support other software kernels. Because it began with support for Julia, Python, and R, it was renamed as JuPyteR, though now the platform supports Scala, Haskell, and Ruby, amongst many others. A list of all kernels supported can be found at https://github.com/jupyter/jupyter/wiki/Jupyter-kernels. Henceforth, you will use Jupyter as your main interface for interactive programming.

Opinionated Docker Stacks

Jupyter provides opinionated stacks for use in a variety of contexts: the All Spark stack, the minimal stack, etc. We have already been using the scipy stack. Let's take a look at the demo stack. The demo stack can be tried immediately by visiting https://try.jupyter.org and is a fully-functioning jupyter system with multiple kernels (Bash, Haskell, Julia, Python 2 and 3, R, Ruby, and Scala) preinstalled.

Security in the Jupyter Notebook Server

One core best practice defined by the Jupyter Docker images maintained by the Jupyter Team is the generation of a security token at container runtime. As part of the run process, a new Jupyter container will generate and write to logs a token that must be passed as a query parameter or entered into the password field in order to interact with the system. This is only necessary the first time that Jupyter is loaded in the browser. The significance of this is that a Jupyter Noteook Server is safe from intruders should we wish to host it on the open web (as described in Chapter 1).

Listing 3-18 demonstrates this by running the jupyter/scipy-notebook image in a new container on port 8888.

Listing 3-18. Run the jupyter/scipy-notebook Image

```
$ docker run -p 8888:8888 jupyter/scipy-notebook
...
 [I 18:53:31.565 NotebookApp] The Jupyter Notebook is running at: http://
[all ip addresses on your system]:8888/?token=44dab68c1bc7b1662041853573f37
cfa03f13d029d397816
[I 18:53:31.565 NotebookApp] Use Control-C to stop this server and shut down
all kernels (twice to skip confirmation).
[C 18:53:31.566 NotebookApp]

    Copy/paste this URL into your browser when you connect for the first time,
    to login with a token:
    http://localhost:8888/?token=44dab68c1bc7b1662041853573f37cfa03f13d02
    9d397816
```

The logs have displayed the token and generated an URL that can be copied and pasted into a browser. You can access the Jupyter Notebook Server by visiting either localhost (Listing 3-19) or the IP identified when you ran docker-machine followed by port 8888 (Listing 3-20).

Listing 3-19. Visiting Jupyter on localhost port 8888 with Provided Token

```
http://localhost:8888/?token=44dab68c1bc7b1662041853573f37cfa03f13d02
9d397816
```

Listing 3-20. Visiting Jupyter on an IP at port 8888 with Provided Token

```
http://192.168.99.100:8888/?token=44dab68c1bc7b1662041853573f37cfa03f13d02
9d397816
```

■ **Note**   The URL generated in the logs will always list the base URL as http://
localhost:8888 even if the server is being run on a different IP and using a different port.
You must modify the URL to reflect the proper IP and port in order to access your systems.

Jupyter Demo Stack

Let's bring the Jupyter Demo Stack online in your local environment. If you are running Docker for Linux, Docker for Mac, or Docker for Windows, you will be able to access any **exposed** Docker containers by visiting http://localhost in a browser at the appropriate port. If you have been running the Docker daemon on a virtual machine (i.e. if you are running Docker Toolbox on Mac or Windows), you will need to get the IP of the virtual host and access your machine there instead. Typically, your Docker virtual machine will be named default. You can find this IP by using the docker-machine ip (see Listing 3-21). Just to reiterate, this is not necessary when using Docker for Linux, Docker for Mac, or Docker for Windows.

Table 3-1 summarizes how you will access your system.

Table 3-1. *Accessing Your Docker System*

OS	Docker System	Shell	Access Jupyter at
Linux	Docker for Linux	Bash	localhost:8888
MacOS >= 10.10.3 (Yosemite)	Docker for Mac	Bash	localhost:8888
MacOS >= 10.8 (Mountain Lion)	Docker Toolbox for Max	Docker Quickstart Terminal	#DOCKERIP:8888
Windows 10 Pro, Enterprise, or Education	Docker for Windows	Windows PowerShell	localhost:8888
Windows 7, 8, 8.1, or 10 Home	Docker Toolbox for Windows	Docker Quickstart Terminal	#DOCKERIP:8888

Listing 3-21. Get the IP of Your Host Machine

```
$ docker-machine ip default
192.168.99.100
```

This signifies that the Docker host virtual machine is available at 192.168.99.100. Your machine may be available at a different IP address.

■ **Note** This command only needs to be run by host systems using Docker Toolbox. Systems running Docker for Linux, Docker for Mac, or Docker for Windows will be able to access their Jupyter Notebook server in browser using localhost.

Launch the jupyter/demo image

To launch an image, you will first pull it from Docker Hub (see Listing 3-22). You could have done this implicitly via the run command but running it explicitly gives greater insight into what you are doing.

Listing 3-22. Pull the Jupyter Demo from Docker Hub

```
$ docker pull jupyter/demo
Using default tag: latest
latest: Pulling from jupyter/demo
...
Digest: sha256:d3dd87e52ca1edbfc8b65ad68bfa91f15eb0660d218c64fd5cdb039c1
fa10818
Status: Downloaded newer image for jupyter/demo:latest
```

Having pulled the image, you run the image using the docker run command (see Listing 3-23).

Listing 3-23. Run the Jupyter Demo

```
$ docker run -p 8888:8888 jupyter/demo
[I 02:54:37.454 NotebookApp] Writing notebook server cookie secret to /home/
jovyan/.local/share/jupyter/runtime/notebook_cookie_secret
[W 02:54:37.537 NotebookApp] WARNING: The notebook server is listening on
all IP addresses and not using encryption. This is not recommended.
[I 02:54:37.580 NotebookApp] Serving notebooks from local directory: /home/
jovyan/work
[I 02:54:37.580 NotebookApp] 0 active kernels
[I 02:54:37.580 NotebookApp] The Jupyter Notebook is running at: http://[all
ip addresses on your system]:8888/
[I 02:54:37.580 NotebookApp] Use Control-C to stop this server and shut down
all kernels (twice to skip confirmation).
```

The vast majority of the output in Listing 3-23 is output generated by the Jupyter application and written to the standard logger. As you are running the container in **foreground** mode, the output is written to your terminal.

You may recall that earlier you attached your working directory to the directory /home/jovyan. You can see in the output here that this is the directory from which Jupyter is serving files:

```
[I 02:54:37.580 NotebookApp] Serving notebooks from local directory: /home/
jovyan/work
```

■ **Note**　Some legacy Jupyter images serve notebooks from the local directory, /home/ jovyan/work. More recent Jupyter images serve notebooks from the local directory, /home/ jovyan/. This will have some impact on your work in terms of the folder used to mount a local directory to a directory within the Jupyter container. If you always use -v 'pwd':/ home/jovyan, then there should be no issues with accessing files or file persistence. It is worth emphasizing that *local directory* in this context refers to the local directory within the running Docker container.

The Jupyter File System

Visiting the application in your browser, you first see the main Jupyter File System (Figure 3-1).

Figure 3-1. *The Jupyter file system*

You can launch a Jupyter file by clicking any of the files in the file system or launch a new file via the "New" menu in the upper right corner.

Open the "Welcome to Python" file by clicking it. After a brief warning about using the file on the free hosting on Rackspace (which you can ignore because you are hosting it yourself), you see the barest of instructions on executing code in a Jupyter file in the first markdown cell (Listing 3-24, Figure 3-2).

Listing 3-24. Run Some Python Code!

```
To run the code below:
Click on the cell to select it.
Press SHIFT+ENTER on your keyboard ...
```

Run some Python code!

To run the code below:

1. Click on the cell to select it.
2. Press SHIFT+ENTER on your keyboard or press the play button (▶) in the toolbar above.

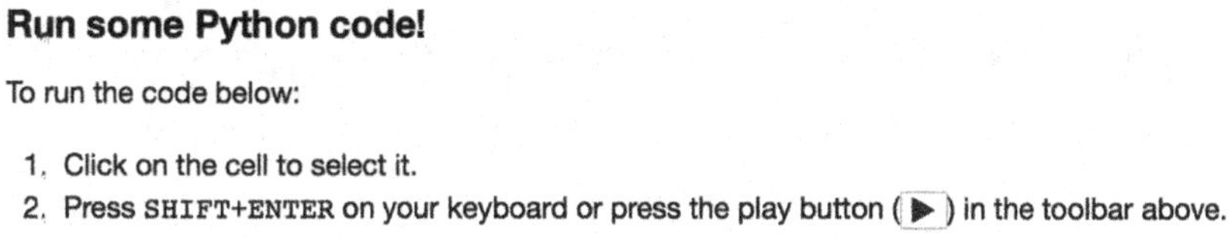

Figure 3-2. *Minimal instructions*

The next cell contains a block of Python (Listing 3-25), which you can execute by pressing Shift+Enter as instructed. The code generates a time series plot of random data (Figure 3-3). Figure 3-4 shows the entire code and result in a Jupyter Notebook.

Listing 3-25. Introduction to Python

```
In [1]: %matplotlib notebook
    import pandas as pd
    import numpy as np
    import matplotlib

    from matplotlib import pyplot as plt
    import seaborn as sns

    ts = pd.Series(np.random.randn(1000), index=pd.date_range('1/1/2000',
    periods=1000))
    ts = ts.cumsum()

    df = pd.DataFrame(np.random.randn(1000, 4), index=ts.index,
    columns=['A', 'B', 'C', 'D'])
    df = df.cumsum()
    df.plot(); plt.legend(loc='best')
Out[1]: <matplotlib.legend.Legend at 0x7f73cc716b38>
```

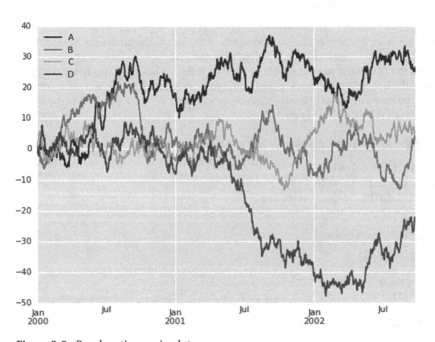

Figure 3-3. *Random time-series data*

Run some Python code!

To run the code below:

1. Click on the cell to select it.
2. Press SHIFT+ENTER on your keyboard or press the play button (▶) in the toolbar above.

A full tutorial for using the notebook interface is available here.

In [1]:
```
%matplotlib notebook

import pandas as pd
import numpy as np
import matplotlib

from matplotlib import pyplot as plt
import seaborn as sns

ts = pd.Series(np.random.randn(1000), index=pd.date_range('1/1/2000', periods=1000))
ts = ts.cumsum()

df = pd.DataFrame(np.random.randn(1000, 4), index=ts.index,
                  columns=['A', 'B', 'C', 'D'])
df = df.cumsum()
df.plot(); plt.legend(loc='best')
```

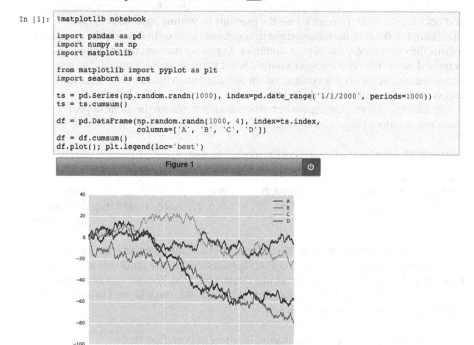

Out[1]: <matplotlib.legend.Legend at 0x7f93c71ce358>

Figure 3-4. *Try a Jupyter Python demo*

Close the Jupyter file by selecting File ➤ Close and Halt. This will terminate the kernel, an IPython process connected to your in-browser notebook, and close the browser window.

You can shut down the Jupyter application by pressing Ctrl+C twice. This will terminate the jupyter notebook process being managed by Docker and thus shut the container down.

```
^C[I 03:10:32.463 NotebookApp] Interrupted...
[I 03:10:32.464 NotebookApp] Shutting down kernels
```

Port Connections

Let's take a brief look at ports. You will also launch your container in **detached** mode via the -d flag (Listing 3-26).

Listing 3-26. Run the jupyter/demo Image in Detached Mode

```
$ docker run -d -p 5000:8888 jupyter/demo
2040f677ad7ffa4666d0d9826e00175a15315ae2b2422314924f6022d6b65622
```

In Listing 3-15, you run the jupyter/demo image in detached mode, exposing port 8888 to port 5000. You can access the machine by visiting localhost:5000 or 192.168.99.100:5000. Because you ran the container in detached mode, the command returns the <container_id> of your container. As you are running the container in **detached** mode, the Jupyter logger's output is not written to your terminal. You can access the output via the logs command. In order to obtain the randomly generated access token, you will need to do just this.

In Listing 3-27, you use the docker ps command to obtain the name of your container in order to access its logs.

Listing 3-27. Display the Containers Currently Running

```
$ docker ps
CONTAINER ID   IMAGE          COMMAND     PORTS                     NAMES
2040f677ad7f   jupyter/demo   "tini ..."  0.0.0.0:5000-\>8888/tcp   furious_
                                                                    archimedes
```

Note that in addition to assigning the container a <container_id>, the Docker daemon also assigned a name, furious_archimedes. Names are randomly assigned, typically an adjective and the surname of a famous scientist. You can use this name to access the generated logs for your container (Listing 3-28). You will need to access the logs in order to obtain your access token.

Listing 3-28. Display Logs for a Container Running in Detached Mode

```
$ docker logs furious_archimedes
[I 15:38:05.402 NotebookApp] Writing notebook server cookie secret to /home/
jovyan/.local/share/jupyter/runtime/notebook_cookie_secret
[W 15:38:05.455 NotebookApp] WARNING: The notebook server is listening on
all IP addresses and not using encryption. This is not recommended.
[I 15:38:05.471 NotebookApp] Serving notebooks from local directory: /home/
jovyan/work
[I 15:38:05.471 NotebookApp] 0 active kernels
[I 15:38:05.471 NotebookApp] The Jupyter Notebook is running at: http://[all
ip addresses on your system]:8888/?token=a7ae5855be48acdb99d12f06f03354cc0b
ede5a941f10d22
[I 15:38:05.472 NotebookApp] Use Control-C to stop this server and shut down
all kernels (twice to skip confirmation).
[C 15:38:05.472 NotebookApp]

    Copy/paste this URL into your browser when you connect for the first
    time, to login with a token:
    http://localhost:8888/?token=a7ae5855be48acdb99d12f06f03354cc0bede5a941
    f10d22
...
```

Port Mappings

You can look at the port mappings for your container with the docker port command (Listing 3-29). This signifies that port 8888 on your container is mapped via the TCP protocol to port 5000 on the host machine (0.0.0.0). If you are running Docker on a virtual machine (i.e. if you are using Docker Toolbox), then access to this port will be at localhost **with respect to the container**. This is to say that it will have an IP on your **system**. You identified this IP earlier via docker-machine ip default where *default* was the name you assigned to your virtual machine. From the persepective of the Docker daemon you are accessing the container's 8888 port via localhost:5000, but if you are running a virtual machine you are accessing it at 192.168.99.100:5000 (Figure 3-5).

Listing 3-29. Examine the Port of a Running Container

```
$ docker port furious_archimedes
8888/tcp -\> 0.0.0.0:5000
```

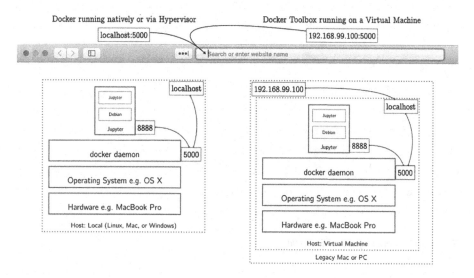

Figure 3-5. *Accessing Jupyter via* localhost

Data Persistence in Docker

Let's consider persistence in Docker. To do this, you will do a quick exercise in Jupyter. The content is fairly basic and is used here solely to demonstrate persistence. Visit the Jupyter application in your browser. Once there, create a new Python file and run a basic calculation on that file. Listing 3-30 is a simple snippet of code for generating a plot for a basic quadratic curve (Figure 3-6). Figure 3-7 shows the entire code and result in a Jupyter Notebook.

Listing 3-30. Plot a Basic Quadratic

```
In [1]: %matplotlib notebook
        import numpy as np
        import matplotlib.pyplot as plt
In [2]: x = np.arange(1,10,1)
        f = lambda x: x**2
        y = f(x)
In [3]: plt.plot(x,y)

Out[3]: <matplotlib.lines.Line2D at 0x7f73cc269978>
```

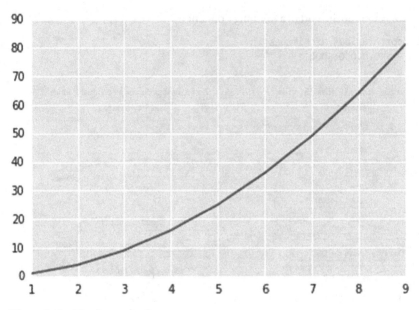

Figure 3-6. *A basic quadratic curve*

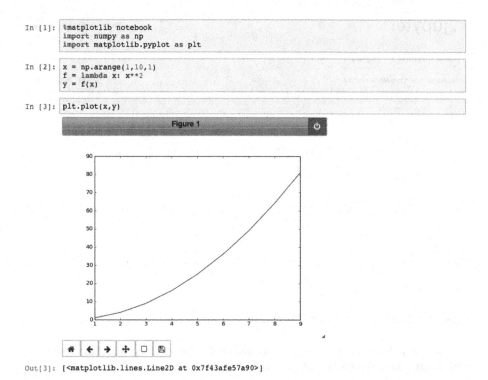

```
In [1]:  %matplotlib notebook
         import numpy as np
         import matplotlib.pyplot as plt
```

```
In [2]:  x = np.arange(1,10,1)
         f = lambda x: x**2
         y = f(x)
```

```
In [3]:  plt.plot(x,y)
```

Out[3]: [<matplotlib.lines.Line2D at 0x7f43afe57a90>]

Figure 3-7. *Generating a basic quadratic curve in a notebook*

Rename the file as Basic Quadratic, save your changes, then choose "Close and Halt" from the file menu. Returning to the Jupyter file system, you should see the file you just created (Figure 3-8).

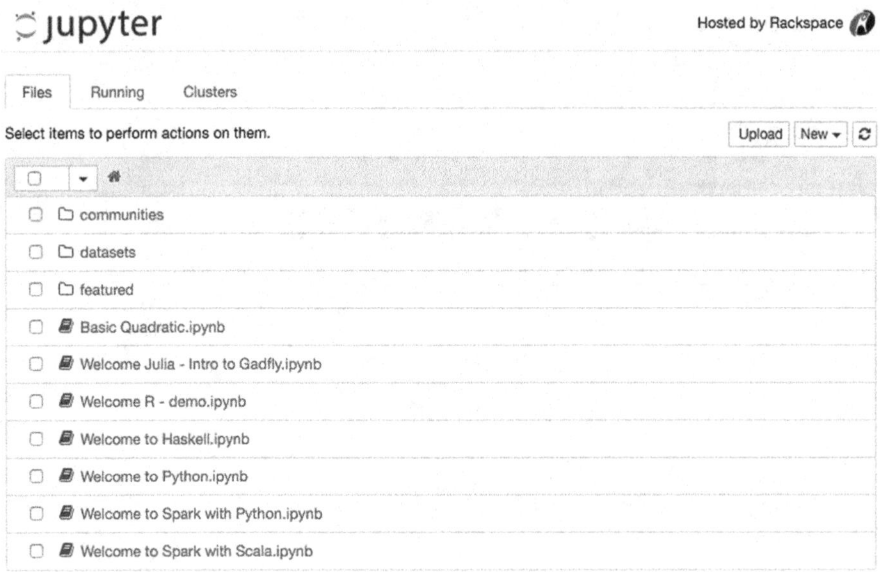

⟳ Jupyter

Files	Running	Clusters

Select items to perform actions on them.   Upload   New ▾   ⟳

☐	▾	🏠
☐	🗁 communities
☐	🗁 datasets
☐	🗁 featured
☐	🕮 Basic Quadratic.ipynb
☐	🕮 Welcome Julia - Intro to Gadfly.ipynb
☐	🕮 Welcome R - demo.ipynb
☐	🕮 Welcome to Haskell.ipynb
☐	🕮 Welcome to Python.ipynb
☐	🕮 Welcome to Spark with Python.ipynb
☐	🕮 Welcome to Spark with Scala.ipynb

Figure 3-8. *Your file in the Jupyter file system*

Next, shut down your running instance and confirm that it is no longer running by using docker stop and docker ps (Listing 3-31).

Listing 3-31. Shut Down a Running Container

```
$ docker stop furious_archimedes
furious_archimedes
$ docker ps
CONTAINER ID     IMAGE      COMMAND      CREATED      STATUS      PORTS      NAMES
```

Now, in Listing 3-32, you start a new Jupyter image. You will need to access the token once more. Here, you view the logs using the returned container id rather than the container's name.

Listing 3-32. Run the jupyter/demo Image in Detached Mode

```
$ docker run -d -p 5000:8888 jupyter/demo
5999d158488d410ac5fbf3a646e4a962d307e968d3cd2f53e60e0a0c7bbe262c
$ docker logs
5999d158488d410ac5fbf3a646e4a962d307e968d3cd2f53e60e0a0c7bbe262c
```

Visit the Jupyter application in your browser. The file you just created is gone. Data has not persisted from instance to instance. This is a problem.

Attach a Volume

This problem can be solved via a run argument. You can attach a volume via a run argument with the flag -v. Note that if you are running Docker on a virtual machine, Mac or Windows, you are only able to mount volumes from /Users (OS X) or C:\Users (Windows), and volumes may need to be made explicitly available to the docker daemon.

You pass the -v flag a single argument that consists of <local_dir>:<container_id>. The Jupyter Demo stack is serving files from /home/jovyan. You will serve files in ~/src to /home/jovyan/src. Note that you must use the absolute path (i.e. /Users/joshuacook/src).

In Listing 3-33, you run in detached mode and attach a volume.

Listing 3-33. Run an Image and Attach a Volume

```
$ docker run \
  -v /Users/joshuacook/src:/home/jovyan/src \
  -d -p 5000:8888 \
  jupyter/demo
273ff71c6755670e21accd197461dd4256fbeb129393d137733f36bcb5432a55
```

Repeat the above experiment and create a new file called Basic Quadratic. You should notice three things.

- All files in ~/src should be immediately available to your Jupyter application.

- Any files that are written into the src directory on the containerized Jupyter application should be written to ~/src on your machine.

- Thus these files should persist from launch to launch.

It is worth repeating this here: if you wish to persist the work that you have done while running Jupyter via Docker, the best practice is to use the -v flag at runtime to mount a local directory to your container.

Summary

In this chapter you did quite a bit. You explored to some extent the nature of non-interactive and interactive programming using Docker. You explored the running of various Jupyter Docker images on your system for the purposes of interactive programming. You detailed the Jupyter Team-defined best practice in notebook security and how to access your Jupyter Notebook Server in a browser. Finally, you briefly explored how port mappings and file persistence can affect your work when using Docker to run Jupyter. I hope that, after this chapter, you are comfortable running Jupyter on your system using Docker, especially in terms of

1. Identifying your security token

2. Identifying your IP and port

3. Persisting the work you do beyond the lifespan of a container

You will revisit Jupyter in Chapter 7, when you explore in some depth the publicly available Docker images written by the Jupyter team for the purposes of running and extending Jupyter via Docker.

## CHAPTER 4

The Docker Engine

The Docker engine is the core technology upon which you will do your work. For your purposes, you can think of the Docker engine as the Docker daemon and the Docker client you use to give the daemon instructions. Docker as a whole consists of both the engine and the Hub, the latter of which is used to store images. If I have not emphasized this enough, the magic happens because we can count on the Docker engine to work the same way no matter our underlying hardware (or virtual hardware) and operating system. We build it using the Docker engine, we test it using the Docker engine, and we deploy it using the Docker engine.

Examining the Docker Workstation

Docker is a rapidly changing technology. As such, it is best to refer to Docker's latest instructions on installation. The base installation includes each of the core technologies we will be using: `docker`, `docker-machine`, and `docker-compose`.

Running the command docker alone returns usage. If you are ever in doubt as to which commands can be run and which arguments they require, try running the command with no arguments (see Listing 4-1).

Listing 4-1. Display Docker Usage

```
$ docker

Usage:  docker COMMAND

A self-sufficient runtime for containers
...
```

You can display system-wide information using the docker info command (Listing 4-2). Either docker or docker info can also function as a minimal verification of a working Docker installation.

© Joshua Cook 2017

J. Cook, *Docker for Data Science*, DOI 10.1007/978-1-4842-3012-1_4

Listing 4-2. Display Docker System Info

```
$ docker info
Containers: 12
 Running: 2
 Paused: 0
 Stopped: 10
Images: 23
Server Version: 17.06.0-ce
...
```

In Listing 4-3, you pull the minimal Docker image, `alpine`, to your local collection of images. `alpine` is

> *A minimal Docker image based on Alpine Linux with a complete package index and only 5MB in size!*

and provides an excellent starting point for building minimal images.

Listing 4-3. Pull the `alpine` Docker Image

```
$ docker pull alpine
Using default tag: latest
latest: Pulling from library/alpine
0a8490d0dfd3: Pull complete
Digest: sha256:dfbd4a3a8ebca874ebd2474f044a0b33600d4523d03b0df76e5c5986cb02d7e8
Status: Downloaded newer image for alpine:latest
```

When you pulled the `hello-world` or `ubuntu` image, you did so implicitly as part of the `docker run` command (see Figure 4-1). In Listing 4-3, you explicitly pull the image. Since you did not specify a tag for pulling, the Docker engine defaults to using the tag named `latest`. The Docker engine then finds the latest `alpine` image by name on Docker Hub and downloads the image to a local image cache (see Figure 4-1). The `alpine` image consists of a single layer. Had it consisted of more than one layer, they would have been pulled in parallel.

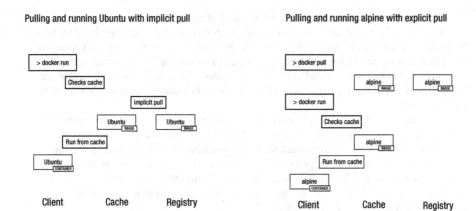

Figure 4-1. *Implicit vs. explicit pulls of images from the Docker Registry*

If you wish to run an interactive shell to the alpine image you downloaded, you can do so via the run command (see Listing 4-4). You will receive a shell prompt for a shell attached to the alpine image. Press Ctrl+D to terminate the shell process and return to the host system.

Listing 4-4. Run an Interactive Shell to an alpine Container

```
$ docker run -it alpine /bin/sh
/ # ls
bin    etc   lib    mnt    root   sbin   sys   usr
dev    home  media  proc   run    srv    tmp   var
/ # whoami
root
/ # ps
PID   USER    TIME    COMMAND
1     root    0:00    /bin/sh
7     root    0:00    ps
/ # ps aux
PID   USER    TIME    COMMAND
1     root    0:00    /bin/sh
8     root    0:00    ps aux
```

In Listing 4-4, the -i flag starts an interactive container. The -t flag allocates a pseudo-TTY command line interpreter or shell that attaches stdin and stdout. You ran the ls command and displayed the directories in the root directory ('/') of the running Docker container. You then ran ps and ps aux to display processes attached to the current shell and all process, respectively.

What exactly have you done? Well, it's somewhat abstract. You have launched an instance of the alpine image. The Docker engine then created a temporary layer on top of this image that you can interact with. The underlying image and this read/write layer on top of that image comprise a Docker container. The Docker engine then connected you to the container via a generic shell. Here you have root access to that running container. The results of any actions you conduct while connected via the shell are written to the top read/write layer.

As discussed in Chapter 2, a Docker container can be thought of as a virtualized process. We will continue to use this way of thinking, and I will use the terms *process* and *container* somewhat interchangeably. So far, you have used docker ps to display processes currently being managed by the Docker daemon (that is, the currently running containers). In Listing 4-5, you display all processes/containers, including those that have terminated via docker ps -a. Here, the -a flag signifies that **all** containers should be shown, including those that have exited.

Listing 4-5. Display All Docker Processes

```
$ docker ps -a
CONTAINER ID  IMAGE   COMMAND    CREATED        STATUS     PORTS    NAMES
b04dfee8fc1c  alpine  "/bin/sh"  49 minutes ago Exited ... clever_khorana
```

Here you see the process you recently terminated. The container created to run the process has the container id b04dfee8fc1c.

When you terminate the shell, the Docker engine stops the container. The container defined by the alpine image and the read/write layer created at runtime continue to exist in your cache. The read/write layer, however, is associated with *this* container, not the underlying image. If you launch a new container by running the same command again, the Docker engine will create a new read/write layer on top of the base image as a new container.

In Listing 4-6, you create a new container, shut down, and list all processes once more. You can see that you now have two containers that have exited associated with the alpine image.

Listing 4-6. Create a Second alpine Container, Terminate, and Display All Processes

```
$ docker run -it alpine /bin/sh
/ #^d
$ docker ps -a
CONTAINER ID  IMAGE   COMMAND    CREATED        STATUS     PORTS    NAMES
b04dfee8fc1c  alpine  "/bin/sh"  49 minutes ago Exited ... clever_khorana
4d3cc1c5471d  alpine  "/bin/sh"  59 minutes ago Exited ... suspicious_ritchie
```

Hello, World in a Container

docker run is used to run an application as a container. Let's walk through this. Let's begin by displaying the currently running docker processes (see Listing 4-7).

Listing 4-7. Display the Docker Processes

```
$ docker ps
CONTAINER ID    IMAGE    COMMAND    CREATED    STATUS    PORTS    NAMES
```

You should see the headers of an empty table signifying that there are no containers/processes currently running. You may see other processes there and that is okay for your purposes.

You can view the images that you currently have in your image cache. You can do this via the docker images command, as shown in Listing 4-8.

Listing 4-8. Display Docker Images

```
$ docker images
REPOSITORY      TAG       IMAGE ID        CREATED        SIZE
alpine          latest    88e169ea8f46    7 weeks ago    3.98 MB
ubuntu          latest    b549a9959a66    31 hours ago   188 MB
hello-world     latest    690ed74de00f    4 weeks ago    960 B
```

If you have been following along, you should have three images. The significance of these three images is that they are stored locally. You can run a container defined by any one of them instantly; that is to say, the image does not need to be implicitly pulled before the container is run. In Listing 4-9, you run the hello-world image.

Listing 4-9. Run the hello-world Image

```
$ docker run hello-world
Hello from Docker.
This message shows that your installation appears to be working correctly.
```

To generate this message, Docker took the following steps:

1. The Docker client contacted the Docker daemon.

2. The Docker daemon pulled the "hello-world" image from the Docker Hub.

3. The Docker daemon created a new container from that image which runs the executable that produces the output you are currently reading.

4. The Docker daemon streamed that output to the Docker client, which sent it to your terminal.

Previously, you ran this command and had to wait while the images were pulled from the registry. Now it runs immediately and shows the same response.

In Listing 4-10, you display processes currently being managed by Docker; again you should see none of the processes you have been working with in this chapter. The significance of this is that the hello-world image launched, displayed its message, and then shut down again.

Listing 4-10. Display Currently Running Containers/Processes

```
$ docker ps
CONTAINER ID     IMAGE      COMMAND      CREATED      STATUS      PORTS      NAMES
```

Run Echo as a Service

Next, let's echo the phrase "Hello, World!" as a service. In the Docker ecosystem, a service is a process that has been "containerized." As you may have noticed, I have been referring to the results of a docker run command as a container/process. This idea is critical to what we are attempting to do. We are wrapping a process *in* a container, like "Hello World!" as you pass a command to be executed by your container (Listing 4-11).

Listing 4-11. Echo "Hello, World!" as a Service

```
$ docker run alpine /bin/echo 'Hello, World!'
Hello World!
```

Again, you display currently running containers/processes to confirm that the alpine container shut down again (Listing 4-12). Again, none of the processes with which you have been working are running. The alpine image launched, used the echo command to echo "Hello World!," and then shut down.

Listing 4-12. Display Currently Running Containers/Processes

```
$ docker ps
CONTAINER ID     IMAGE      COMMAND      CREATED      STATUS      PORTS      NAMES
```

The gravity of this may not be apparent. To fully comprehend what is happening, let's time the whole thing. Let's use the built-in time command to see how long it takes to echo a string as a service. In Listing 4-13, you do this by running time and then the command you ran previously.

Listing 4-13. Time the Execution of the Echo Service

```
$ time docker run alpine /bin/echo 'Hello World!'
Hello World!

real    0m1.300s
user    0m0.019s
sys     0m0.030s
```

Now you use the built-in time command to see how long it takes to do the same natively (that is, directly on your machine); see Listing 4-14.

Listing 4-14. Time the Execution of echo Natively

```
$ time echo 'Hello World!'
Hello World!

real   0m0.006s
user   0m0.000s
sys    0m0.001s
```

That is a pretty significant jump. Running the command in the Docker instance made the task take 200 times longer. But consider this from another perspective. Instead, consider the fact that booting up an entire alpine instance only added about a second to your processing time.

Isolating the Bootstrap Time

In Listing 4-15, you attempt to isolate the bootstrap time. First, you run and time a hard sleep, again with the time command. The sleep command simply puts the shell to sleep for a specified amount of time. This will allow you to explicitly control timing.

Listing 4-15. Time a Hard Sleep Natively

```
$ time sleep 2

real   0m2.012s
user   0m0.001s
sys    0m0.003s
```

In Listing 4-16, you run this hard sleep as a service and time the execution. It is useful to spend a moment to look at how this is done. In Listing 4-16, you run time docker run alpine /bin/sleep 2. You are timing the use of docker to manage a sleep of 2 seconds run in a container defined by the alpine image.

Listing 4-16. Time a Hard Sleep as a Service

```
$ time docker run alpine /bin/sleep 2

real   0m3.061s
user   0m0.014s
sys    0m0.013s
```

Note that you see approximately the same increase. In your rudimentary testing, you can take this to signify that "containerizing" your service adds about a second to your runtime. This is significant when echoing "Hello World!" or performing a hard sleep of 2 seconds. This is completely negligible when running a jupyter server or performing a long calculation with numpy.

A Daemonized Hello World

You will ultimately want to daemonize your Docker containers (that is, set them up to run indefinitely as background processes). You do this by running your containers in detached mode via the -d flag.

Let's run alpine in detached mode (that is, run the image as a container, and leave it running in the background). In order to keep it running, you will give it a job to do, namely echo "hello world" every second ad infinitum. Alternatively, you could give it a more complex job to do, such as a long arithmetic operation or to listen for web requests, but for demonstration purposes, this will suffice (Listing 4-17).

Listing 4-17. Run the Alpine Image in Detached Mode

```
$ docker run -d alpine /bin/sh -c "while true; do echo hello world; sleep 1; done"
```

In Listing 4-18, you again display the running containers/processes. Because you passed the run command the -d flag for detached mode, it is still running.

Listing 4-18. Display Currently Running Containers/Processes

```
$ docker ps
CONTAINER ID  IMAGE   COMMAND      CREATED        STATUS  PORTS  NAMES
9201755545d1  alpine  "/bin/sh ..."  51 seconds ago  Up ...         upbeat_easley
```

But how do you confirm that it is doing the job it has been tasked with? You can do this via the command docker logs. In Listing 4-19, you request the logs associate with the alpine container named upbeat_easley.

Listing 4-19. Show Logs for a Detached Container

```
$ docker logs upbeat_easley
hello world
hello world
...
```

Here, you view logs for your container currently running in detached mode. Not the most exciting log, but sufficient to confirm that your container is doing its job.

Let's give it a rest, using the stop command to shut it down. In Listing 4-20, you stop the alpine container upbeat_easley.

Listing 4-20. Stop a Detached Container

```
$ docker stop upbeat_easley
```

Summary

This was chapter was the shortest thus far in the text. In it, you explored a few aspects of asking Docker to manage processes as Docker containers. The substance of this chapter is somewhat abstract. A thorough understanding of the material is not, strictly speaking, necessary for the mastery of the big-picture concepts in this book. Readers who are interested in digging deeper into the nuances of the Docker engine are encouraged to peruse the excellent and in-depth documentations available at `http://docs.docker.com`.

■ ■ ■

## CHAPTER 5

The Dockerfile

Every Docker image is defined as a stack of layers, each defining fundamental, stateless changes to the image. The first layer might be the virtual machine's operating system (a Debian or Ubuntu Docker image), the next the installation of dependencies necessary for your application to run, and all the way up to the source code of your application. The best way to leverage this system is via a Dockerfile.

Best Practices

A Dockerfile is a way to script the building of an image. This script uses a domain-specific language to tell the Docker daemon how to sequentially build a Docker image. When the daemon is instructed to build an image, it does so by reading the necessary instructions from a Dockerfile.

Docker holds the following as best practices when creating Dockerfiles:

- Containers should be ephemeral or stateless, in that they can be reinstantiated with a minimum of set up and configuration.

- Use a .dockerignore file, similar to a .gitignore.

- Avoid installing unnecessary packages.

- Each container should have only one concern.

- Minimize the number of layers.

- Sort multiline arguments.

Stateless Containers

A best practice in modern software development is software that runs as isolated processes sharing nothing between them, often referred to as microservices.[1] While an application may run more than one process at a time, these processes should be stateless. If any information is to be persisted beyond the termination of the process, this information should be written to a stateful backing service such as a database (like MongoDB or Postgres) or a key-value store (like Redis).

---

[1]https://en.wikipedia.org/wiki/Microservices

© Joshua Cook 2017
J. Cook, *Docker for Data Science*, DOI 10.1007/978-1-4842-3012-1_5

We look at Docker containers as abstractions of system processes, in fact, thinking of Docker containers as processes being managed by the Docker daemon. As such, it is also best practice to define our Docker images, and thus the running containers which they define, as completely stateless. We should be able to shut down a container and remove it from our system, then start an identical container using the same image, and run command with **no effect to our work**.

Single-Concern Containers

Early iterations of Dockerfile best practices held to the mantra "one process per container." As Docker has matured, sticking to one process for each container has proven untenable. Jupyter is actually a prime example of a container that can't be defined to run via a single process.

That said, best practice holds that each container should be defined to have a single *concern*. Jupyter may require multiple processes to function properly but we will dedicate a single container to running Jupyter. Should we wish to interface with a Postgres database, we would use a second container concerned with this database. The end goal is to keep application concerns separate and modular.

Project: A Repo of Docker Images

In this chapter, you will be developing a repository of Dockerfiles, each defining a separate image you might use in the course of your work. In the process of developing and maintaining these images defined via a Dockerfile, you will explore the syntax of defining an image and best practices in building and maintaining images. Your ultimate goal is to create a suite of images that will become your primary building blocks in developing modular systems for performing the work of the data scientist.

Prepare for Local Development

First, you will prepare your local machine for the development work you will be doing. In Listing 5-1, you do this by creating a directory for your project and initializing it as a git repository.

Listing 5-1. Prepare for Local Development

```
$ mkdir ch_5_dockerfiles && cd ch_5_dockerfiles
$ git init
$ touch README.md
$ git add README.md
$ git commit -m 'init'
```

Configure GitHub

On GitHub, create a repo called Dockerfiles. Once you have created the repo, connect your local directory to the remote GitHub repo (Listing 5-2).

■ **Note**   You will need to configure your GitHub account for connection via SSH. Documentation for this is available at `https://help.github.com/articles/connecting-to-github-with-ssh/`.

Listing 5-2. Connect the Local Repo to GitHub

```
$ git remote add origin git@github.com:<username>/dockerfiles.git
$ git push -u origin master
```

Building Images Using Dockerfiles

The connection between the Dockerfile and a built (compiled) Docker image is the docker build command. The build command tells the Docker daemon to construct an image using the specified context and a Dockerfile. Context refers to the collection of files that will be used to build the specific image. The context will be specified by an included PATH. You can take a look at the requirements of docker build to see what this might look like (Listing 5-3).

Listing 5-3. Display docker build help

```
$ docker build
"docker build" requires exactly 1 argument(s).
See 'docker build --help'.

Usage:  docker build [OPTIONS] PATH | URL | -
```

As you can see, docker build requires a PATH or a URL as the final argument. This is the context.

Dockerfile Syntax

Dockerfiles are built using a simple domain-specific language (see Listing 5-4). Instructions are case-insensitive but by convention are written in all caps. Instructions are passed sequentially and Dockerfiles should be thought of as scripts passed to the Docker daemon.

Listing 5-4. Dockerfile syntax

```
# Comment
INSTRUCTION arguments
```

Designing the `gsl` Image

The first image you will construct is the same image you used in Chapter 3, the GSL image. This image is used for compiling code using the GNU Scientific Library (GSL), a suite of C tools used in computational mathematics, especially using the BLAS ecosystem. You build this image using the gcc[2] image as a base image, and in doing so ensure that you have all of the tools necessary to compile your C code.

Create the `gsl` Source Directory

First, you make a directory for this specific image and instantiate your Dockerfile (Listing 5-5).

Listing 5-5. Create a Directory for the `gsl` Image Containing an Empty Dockerfile

```
$ mkdir gsl
$ touch gsl/Dockerfile
```

In Listing 5-6, to build the gsl image, you use the docker build command, using the relatively-referenced folder gsl as context. The -t flag tells the daemon to name that image joshuacook/gsl. Your initial attempt at a build fails because you have not added any commands to the Dockerfile.

Listing 5-6. Run a Docker Build

```
$ docker build -t joshuacook/gsl gsl
sending build context to Docker daemon 53.25 kB
Error response from daemon: The Dockerfile (Dockerfile) cannot be empty
```

Define the `gsl` Image

Let's define the gsl image as you did in Chapter 3, using three layers, defined by three commands, as seen in Listing 5-7.

Listing 5-7. gsl/Dockerfile

```
FROM gcc

LABEL maintainer=@joshuacook

RUN apt-get update && \
      apt-get install -y \
      gsl-bin \
      libgsl0-dbg \
      libgsl0-dev \
      libgsl0ldbl
```

---

[2]https://en.wikipedia.org/wiki/GNU_Compiler_Collection

Build the gsl Image

In Listing 5-8, having defined the image using a Dockerfile, you build the image using the docker build command, again naming the image and providing the gsl directory as context.

Listing 5-8. Run a Docker Build

```
$ docker build -t joshuacook/gsl gsl
sending build context to Docker daemon 14.85 kB
Step 1 : FROM gcc
latest: Pulling from library/gcc

693502eb7dfb: Pull complete
...
e6a66f7b6a7a: Pull complete
Digest: sha256:c1fa0b3eeba33a7b9da5ab7de7fa2c520760f778b5e5d1db38791d0da7b841b9
Status: Downloaded newer image for gcc:latest
 ---> 408d218617ca
Step 2 : LABEL maintainer @joshuacook
 ---> Running in 723583dd01d6
 ---> 20bbe26b8b8a
Removing intermediate container 723583dd01d6
Step 3 : RUN apt-get update &&    apt-get install -y
gsl-bin    libgsl0-dbg    libgsl0-dev    libgsl0ldbl
 ---> Running in ed02ca384066
Get:1 http://security.debian.org jessie/updates InRelease [63.1 kB]
...
```

FROM gcc

The first layer uses the FROM instruction to define the base image from which you will build your image. A valid Dockerfile must always begin with a FROM instruction. Best practice recommends that all but advanced use cases should begin their images by pulling FROM the Docker Public Repositories. To reiterate, FROM must be the first non-comment instruction in the Dockerfile. Here, you are pulling the from the gcc image.

FROM can be used with the following three syntaxes:

- FROM <image>

- FROM <image>:<tag>

- FROM <image>@<digest>

■ **Note** The use of tag or digest are optional. If neither is provided, the tag is assumed to be latest, corresponding to the latest available build at Docker Hub.

LABEL maintainer=@joshuacook

The second layer uses the LABEL instruction to define metadata associated with your image. Each LABEL is a key-value pair. In this case, you associate the key maintainer (of the image) to the value (the Docker Hub user), joshuacook. Images can have multiple LABELs.

To view an image's LABELs, you use the docker inspect command. In Listing 5-9, you do this for the joshuacook/gsl image that you just built.

Listing 5-9. Inspect the joshuacook/gsl Image

```
$ docker inspect joshuacook/gsl
...
    "Labels": {
        "maintainer": "@joshuacook"
    }
...
```

RUN apt-get update && apt-get install

The final layer of the image uses the RUN instruction, in this case to install the libraries necessary for using the GNU Scientific Library. The RUN instruction uses /bin/sh to execute any provided commands in a new layer on top of the current image. The results of the execution become the new image.

■ **Note**   We use a \ (backslash) to continue a single instruction across multiple lines. Thus, the instructions in Listing 5-10 and the Listing 5-11 are functionally equivalent.

Listing 5-10. apt Install Over Multiple Lines

```
RUN apt-get update && \
    apt-get install -y \
    gsl-bin \
    libgsl0-dbg \
    libgsl0-dev \
    libgsl0ldbl
```

Listing 5-11. apt Install in a Single Line

```
RUN apt-get update && apt-get install -y gsl-bin libgsl0-dbg libgsl0-dev libgsl0ldbl
```

Commit Changes to GitHub

Before you move on, you should git commit the changes that you have made to your Dockerfile and push the changes to GitHub (see Listing 5-12).

Listing 5-12. Commit Changes and Push to GitHub

```
$ git add gsl/Dockerfile
$ git commit -m 'GSL IMAGE - initial build'
$ git push
```

The Docker Build Cache

Recall that the structure of a Docker image is defined as a stack of images, each defining a stateless change to the image. Because each of them is built as a separate abstraction layer, changes to the Docker image are made from the top down when rebuilding the image. If you have only made changes to the code sitting on the top-most layer, this is the only layer of the image that will be rebuilt.

You can examine the build cache by adding a new layer to your GSL image. Let's add a new layer to your image containing another library you might be interested in for performing data science tasks in C. Add the line in Listing 5-13 to your Dockerfile.

Listing 5-13. Add an Additional Dependency Layer to joshuacook/gsl

```
RUN apt-get update && \
    apt-get install -y gdb
```

In Listing 5-14, you rebuild the image.

Listing 5-14. Rerun the joshuacook/gsl Image Build

```
$ docker build -t joshuacook/gsl gsl
Sending build context to Docker daemon 2.048 kB
Step 1/4 : FROM gcc
 ---> 408d218617ca
Step 2/4 : LABEL maintainer @joshuacook
 ---> Using cache
 ---> ba0d6482c28a
Step 3/4 : RUN apt-get update &&        apt-get install -y
gsl-bin       libgsl0-dbg       libgsl0-dev       libgsl0ldbl
 ---> Using cache
 ---> 8706282586fc
Step 4/4 : RUN apt-get install -y gdb
 ---> Running in c491003e5a95

...


 ---> 53111721b96a
Removing intermediate container c491003e5a95
Successfully built 53111721b96a
...
```

■ **Note** During this build process, steps 1-3 were not executed as before. Prior to executing each step, the Docker daemon compares the step to the individual layers of an existing instance of the same image in the local cache. If they match, the build uses the existing layers (i.e. the Docker Build Cache) to build the image. As a result, a build with no changes is idempotent, meaning that the initial command builds the image, but subsequent builds have no additional effect. Listing 5-15 demonstrates the idempotency of the docker build process.

Listing 5-15. Rerun the Identical Build Once More

```
$ docker build -t joshuacook/gsl gsl
Sending build context to Docker daemon 2.048 kB
Step 1/4 : FROM gcc
 ---> 408d218617ca
Step 2/4 : LABEL maintainer @joshuacook
 ---> Using cache
 ---> ba0d6482c28a
Step 3/4 : RUN apt-get update &&        apt-get install -y
gsl-bin        libgsl0-dbg       libgsl0-dev        libgsl0ldbl
 ---> Using cache
 ---> 8706282586fc
Step 4/4 : RUN apt-get install -y gdb
 ---> Using cache
 ---> 53111721b96a
Successfully built 53111721b96a
```

A more interesting and useful result is that a build will only run steps following the first step upon which you have made a change, as demonstrated in Listing 5-15.

Anaconda

Anaconda is a freemium distribution of the Python and R programming languages for large-scale data processing, predictive analytics, and scientific computing. You will use Anaconda to drive your Jupyter platform. The conda command is the primary interface for managing Anaconda installations. Miniconda is a small "bootstrap" version that includes only conda and conda-build, and installs Python. As an exercise, however, you will design your own miniconda image modeled after the miniconda image distributed by Continuum Analytics that can be used to minimally run IPython.

Design the miniconda3 Image

In Chapter 3, you used the jupyter/scipy-notebook image to run IPython using the scipy library. This is a perfectly reasonable usage of that image. It contains all of the libraries that you need. Furthermore, storing two independently defined images (one for the purpose of running Jupyter with scipy and the other for the purpose of running IPython with scipy) would consume unnecessary disk space in the interest of having single-usage images. Finally, running IPython from the `jupyter` image is a recommended best practice from the IPython team.[3] Ultimately, you will do the same.

Create the `miniconda3` Source Directory

Again, in Listing 5-16, you begin by creating a directory to hold the source code associated with this image.

Listing 5-16. Create a Directory for the `miniconda3` Image Containing an Empty Dockerfile

```
$ mkdir miniconda3
$ touch miniconda3/Dockerfile
```

In Listing 5-17, you use the `tree` tool to display the overall structure of your docker images project to this point.

Listing 5-17. Use `tree` to Show Project Progress

```
$ tree
.
├── README.md
├── gsl
│   └── Dockerfile
└── miniconda3
    └── Dockerfile
```

Begin the Image with `FROM`, `ARG`, and `MAINTAINER`

You begin your image with the `FROM` instruction, which sets the **base image** upon which you will build your image. A valid `Dockerfile` requires a `FROM` instruction as its *first* instruction. In Listing 5-18, you take your cues from the Jupyter project and build using the debian image.

Listing 5-18. `miniconda3/Dockerfile`

```
FROM debian
```

---

[3]https://hub.docker.com/r/ipython/ipython/

In Listing 5-19, you build the image.

Listing 5-19. Build the miniconda3 Image

```
$ docker build -t miniconda3 miniconda3
Sending build context to Docker daemon 2.048 kB
Step 1/1 : FROM debian
latest: Pulling from library/debian
6d827a3ef358: Pull complete
Digest: sha256:72f784399fd2719b4cb4e16ef8e369a39dc67f53d978cd3e2e7bf4e502c
7b793
Status: Downloaded newer image for debian:latest
 ---> 8cedef9d7368
Successfully built 8cedef9d7368
```

Here, you have built an image with a single layer: your underlying operating system. In Listing 5-20, you list images in your image cache.

Listing 5-20. Show Cached Images

```
$ docker images
REPOSITORY          TAG        IMAGE ID        CREATED        SIZE
debian              latest     47af6ca8a14a    8 days ago     125.1 MB
miniconda3          latest     47af6ca8a14a    5 minutes ago  125.1 MB
...
```

■ **Note**   The debian image and the miniconda3 image have the exact same IMAGE ID. This is because they **are** the same image.

Commit Changes to the Local Repository

You will be building a fairly complicated image here and as such it is best practice to commit to the local git repository frequently. In Listing 5-21, you make your first commit for this image.

Listing 5-21. Commit Changes

```
$ git add Dockerfile
$ git commit -m 'MINICONDA3 IMAGE. Added FROM instruction.'
```

In Listing 5-22, you add a label for the maintainer of the image, as you did previously for the gsl image. Again, this is not required, but it is conventional.

Listing 5-22. miniconda3/Dockerfile

```
FROM debian
LABEL maintainer=@joshuacook
```

In Listing 5-23, you add an ARG instruction. The ARG instruction is used to define an environment variable that is available at build or runtime.

Listing 5-23. `miniconda3/Dockerfile`

```
FROM debian
LABEL maintainer=@joshuacook
ARG DEBIAN_FRONTEND=noninteractive
```

Here you use the ARG instruction to define an environment variable that will describe the behavior of your interaction with your Debian container. You could also have defined an ARG with no value and then passed the value to the build as an argument (see Listing 5-24).

Listing 5-24. An ARG Passed at Build Time

```
$ DEBIAN_FRONTEND=noninteractive docker build -t some_image .
```

■ **Note** You declare the value of an argument inline with the `docker build` command.

There is yet a third way to declare environment variables. The Jupyter image defined by the Jupyter development team uses the ENV instruction to achieve the same purpose. After reading this thread[4] on the Docker team's GitHub page, I chose to use the ARG instruction. Any of them will work but it is important to only use one.

Idempotently Run the Build

In Listing 5-25, you run the build. As you have previously run the build for your first layer, this will not need to be executed, and the build will pick up only the most recent additions.

Listing 5-25. Run the Build

```
$ docker build -t miniconda3 miniconda3
Sending build context to Docker daemon 2.048 kB
Step 1/3 : FROM debian
 ---> a2ff708b7413
Step 2/3 : LABEL maintainer @joshuacook
 ---> Using cache
 ---> c140313c988e
Step 3/3 : ARG DEBIAN_FRONTEND=noninteractive
 ---> Running in b927c25267f0
 ---> da987fd59d24
Removing intermediate container b927c25267f0
Successfully built da987fd59d24
```

---

[4]https://github.com/docker/docker/issues/4032

Now you're getting somewhere. You have added three lines to your Dockerfile. When you ran the build, it either 1) used a pre-existing layer as in the case at Step 1 or 2) created a new layer containing the results of the step.

■ **Note** The debian image was not pulled this time. There was nothing new to do here!

Commit Changes to the Local Repository

In Listing 5-26, you again commit your changes to the local repository.

Listing 5-26. Commit Changes

```
$ git add Dockerfile
$ git commit -m 'MINICONDA3 IMAGE. Added maintainer LABEL and ARG
instruction'
```

Provision the miniconda3 Image

Ultimately, your Jupyter application will run on a containerized Debian machine. As with any application, it will have many dependencies, or system-level applications, that the Debian system must be able to execute in order to function properly. Were you building this system manually, that is without Docker, you would use the command line package manager appropriate to your operating system (apt for Ubuntu or Debian, yum for CentOS, or brew for Mac OS X) in a process known as provisioning.

The Docker daemon has no native way of provisioning, but rather has a mechanism for leveraging the base system's package manager via the RUN instruction (Listing 5-27).

Listing 5-27. The RUN Instruction

```
RUN <command>
```

This has the effect of running <command> in a shell to the image (i.e. /bin/sh -c <command>). Similar to scripting in a shell language, you can add a backslash (\) to the end of a line to continue that command on the next line. In other words, the statements in Listing 5-28 and Listing 5-29 are equivalent.

Listing 5-28. A RUN Instruction

```
RUN /bin/bash -c 'source $HOME/.bashrc ; echo $HOME'
```

Listing 5-29. Another RUN Instruction

```
RUN /bin/bash -c 'source $HOME/.bashrc ;\
echo $HOME'
```

In Listing 5-30, you provision the thin operating system for the `miniconda3` image.

Listing 5-30. `miniconda3/Dockerfile`

```
RUN apt-get update --fix-missing && \
    apt-get install -y \
    wget bzip2 ca-certificates \
    libglib2.0-0 libxext6 libsm6 libxrender1
```

Run the Build

At this point, you actually have some meat to your image. Let's go ahead and build it (Listing 5-31).

■ **Note** In `Step 4/4`, the RUN instruction is blissfully unaware of those line breaks. They are simply there for us in order to make the `Dockerfile` more readable.

Listing 5-31. Run the Build

```
$ docker build -t miniconda3 miniconda3
Sending build context to Docker daemon  2.56 kB
Step 1/4 : FROM debian
 ---> a2ff708b7413
Step 2/4 : LABEL maintainer @joshuacook
 ---> Using cache
 ---> c140313c988e
Step 3/4 : ARG DEBIAN_FRONTEND=noninteractive
 ---> Using cache
 ---> 6be1e058f2de
Step 4/4 : RUN apt-get update &&     apt-get install -yq --no-
install-recommends     build-essential     bzip2     ca-
certificates     git     libglib2.0-0     libsm6     libxrender1     wget
&& apt-get clean &&     rm -rf /var/lib/apt/lists/*
 ---> Running in 642c7d024e55
Get:1 http://security.debian.org jessie/updates InRelease [63.1 kB]
Ign http://deb.debian.org jessie InRelease
Get:2 http://deb.debian.org jessie-updates InRelease [145 kB]
...
```

This phase of the build will take some time. For now, let's hold off on examining the output. Again, running a Docker build is idempotent, meaning that no matter how many times you run it, you receive the same output. You will wait for this run to complete and then run the build once more to receive a compact output (Listing 5-32).

Listing 5-32. Idempotently Run the Build

```
$ docker build -t miniconda3 miniconda3
Sending build context to Docker daemon  2.56 kB
Step 1/4 : FROM debian
 ---> a2ff708b7413
Step 2/4 : LABEL maintainer @joshuacook
 ---> Using cache
 ---> c140313c988e
Step 3/4 : ARG DEBIAN_FRONTEND=noninteractive
 ---> Using cache
 ---> c140313c988e
Step 4/4 : RUN apt-get update &&    apt-get install -yq --no-
install-recommends    build-essential    bzip2    ca-
certificates    git    libglib2.0-0    libsm6    libxrender1    wget
&& apt-get clean &&    rm -rf /var/lib/apt/lists/*
 ---> Using cache
 ---> bb1ae69e4c8a
Successfully built bb1ae69e4c8a
```

In Listing 5-32, you can see that the build output contains four steps representing the four instruction in your Dockerfile. The way that Docker works, it creates or uses a cached image for each step. At Step 1, you tell the Docker daemon which image you will be using as your base image FROM debian. After each step, you have a new layer and thus a new image and IMAGE ID. Let's have a look at the images in your cache once more (Listing 5-33).

Listing 5-33. Display Images in the Local Image Cache

```
$ docker images
REPOSITORY      TAG       IMAGE ID        CREATED           SIZE
debian          latest    47af6ca8a14a    8 days ago        125.1 MB
miniconda3      latest    bb1ae69e4c8a    About a minute ago 1.636 GB
...
```

■ **Note**   The final layer's ID, bb1ae69e4c8a, matches the IMAGE ID of the miniconda3 image.

Commit Changes to the Local Repository

In Listing 5-34, you again commit your changes to the local repository.

Listing 5-34. Commit Changes

```
$ git add Dockerfile
$ git commit -m 'MINICONDA3 IMAGE. OS provision statement.'
```

Install Miniconda

In Listing 5-35, you install Miniconda via a RUN instruction. You have taken this command directly from the Continuum Analytics Docker image for Miniconda.[5] It varies slightly from the installation of Miniconda for the jupyter/base-notebook image.

Listing 5-35. miniconda3/Dockerfile

```
RUN echo 'export PATH=/opt/conda/bin:$PATH' > /etc/profile.d/conda.sh && \
wget --quiet \
    https://repo.continuum.io/miniconda/Miniconda3-4.3.11-Linux-x86_64.sh -O
    ~/miniconda.sh && \
    /bin/bash ~/miniconda.sh -b -p /opt/conda && \
    rm ~/miniconda.sh
```

Run the Build

In Listing 5-36, you run the build to install Miniconda3.

Listing 5-36. Run the Build

```
$ docker build -t miniconda3 miniconda3
Sending build context to Docker daemon  2.56 kB
...
Step 5/5 : RUN echo 'export PATH=/opt/conda/bin:$PATH' > /etc/profile.d/
conda.sh &&     wget --quiet https://repo.continuum.io/miniconda/Miniconda3-
4.3.11-Linux-x86_64.sh -O ~/miniconda.sh &&     /bin/bash ~/miniconda.sh -b
-p /opt/conda &&     rm ~/miniconda.sh
 ---> Running in 6e3605df5b68
...
Python 3.6.0 :: Continuum Analytics, Inc.
creating default environment...
installation finished.
 ---> 02b8f5d04aeb
Removing intermediate container 6e3605df5b68
Successfully built 02b8f5d04aeb
```

Commit the Changes to the Local Repository

In Listing 5-37, you again commit your changes to the local repository.

Listing 5-37. Commit Changes

```
$ git add Dockerfile
$ git commit -m 'MINICONDA3 IMAGE. Install miniconda3.'
```

---

[5]https://github.com/ContinuumIO/docker-images/blob/master/miniconda3/Dockerfile

tini

I think it a bit beyond of the progress you have made to this point to begin a discussion of the PID 1 "Zombie" process problem at this point. The short of it is this: Docker best practices are that we not run unnecessary processes when running our containers. This includes the typical systemd or sysvinit that would be run by your operating system to handle processes and signals. This can lead to containers that can't be gracefully stopped or zombie containers that persist when they should have died.

For now, let's simply handle this situation the way that the Continuum development team does and use tini. tini is a lightweight solution to this problem with no additional dependencies. It reaps zombies, spawns a single process (which will run as PID 1), and when the tini's first child process has exited, tini exits as well.

In Listing 5-38, you include tini in your container.

Listing 5-38. miniconda3/Dockerfile

```
RUN apt-get install -y curl grep sed dpkg && \
    TINI_VERSION=`curl https://github.com/krallin/tini/releases/latest |
    grep -o "/v.*\"" | sed 's:^..\(.*\).$:\1:'` && \
    curl -L "https://github.com/krallin/tini/releases/download/v${TINI_
    VERSION}/tini_${TINI_VERSION}.deb" > tini.deb && \
    dpkg -i tini.deb && \
    rm tini.deb && \
    apt-get clean
```

Run the Build

In Listing 5-39, you run the build to install tini.

Listing 5-39. Run the Build

```
$ docker build -t miniconda3 miniconda3
Sending build context to Docker daemon  2.56 kB
...
Step 6/6 : RUN apt-get install -y curl grep sed dpkg &&      TINI_
VERSION=`curl https://github.com/krallin/tini/releases/latest | grep -o
"/v.*\"" | sed 's:^..\(.*\).$:\1:'` &&      curl -L "https://github.com/
krallin/tini/releases/download/v${TINI_VERSION}/tini_${TINI_VERSION}.deb" >
tini.deb &&      dpkg -i tini.deb &&      rm tini.deb &&      apt-get clean
 ---> Running in b6fa6f29121c
...
Successfully built ce5eb4345473
```

Commit the Changes to the Local Repository

In Listing 5-40, you again commit your changes to the local repository.

Listing 5-40. Commit Changes

```
$ git add Dockerfile
$ git commit -m 'MINICONDA3 IMAGE. Install tini.'
```

Configure the Environment Variable with ENV

In order to make sure that your container runs properly, you need to configure a few environment variables. In Listing 5-41, you do this with the ENV instruction. You first specify some language specific variables,[6] and then you make sure that the conda binary is in the PATH.

Listing 5-41. miniconda3/Dockerfile

```
ENV LANG=C.UTF-8 LC_ALL=C.UTF-8
ENV PATH /opt/conda/bin:$PATH
```

ENTRYPOINT

Ultimately, you provide an ENTRYPOINT for your container. The ENTRYPOINT instruction specifies the process(es) to be launched when the image is instantiated. In Listing 5-42, you tell Docker to run tini as PID 1.

Listing 5-42. miniconda3/Dockerfile

```
ENTRYPOINT [ "/usr/bin/tini", "--" ]
```

Run the Build

In Listing 5-43, you run the build to set your environment variables and the ENTRYPOINT.

Listing 5-43. Run the Build

```
$ docker build -t miniconda3 miniconda3
Sending build context to Docker daemon 3.072 kB
...
Step 7/9 : ENV LANG C.UTF-8 LC_ALL C.UTF-8
...
Step 8/9 : ENV PATH /opt/conda/bin:$PATH
...
Step 9/9 : ENTRYPOINT /usr/bin/tini --
 ---> Running in 3c3776056958
 ---> 014cc8d97486
Removing intermediate container 3c3776056958
Successfully built 014cc8d97486
Successfully tagged miniconda3:latest
```

---

[6]http://unix.stackexchange.com/questions/87745/what-does-lc-all-c-do

Commit the Changes to the Local Repository

In Listing 5-44, you again commit your changes to the local repository. As this marks the completion of your `miniconda3` image, you also push your changes to GitHub.

Listing 5-44. Commit Changes

```
$ git add Dockerfile
$ git commit -m 'MINICONDA3 IMAGE. Set env and entrypoint.'
$ git push
```

Design the `ipython` Image

Your `miniconda3` image is significantly smaller than most of the other images in the Jupyter stack precisely because it comes with next to nothing preinstalled. If you wish to run `ipython`, you are going to need to install this library. Let's use the `FROM` instruction once more to build an `ipython` image using the `miniconda3` image that you just designed.

Create the `ipython` Source Directory

First, you make a directory for this specific image and instantiate your Dockerfile (Listing 5-45).

Listing 5-45. Create a Directory for the `ipython` Image Containing an Empty `Dockerfile`

```
$ mkdir ipython
$ touch ipython/Dockerfile
```

Define the `ipython` Image

Let's define the `ipython` image, using three layers, defined by three commands, as seen in Listing 5-46.

Listing 5-46. `ipython/Dockerfile`

```
FROM miniconda3
LABEL maintainer=@joshuacook
RUN conda update conda && \
    conda install --quiet --yes ipython && \
    conda clean -tipsy
```

Install `ipython` with `conda`

The only novel command here is using `conda` to install `ipython`. `conda` is a package manager much like apt, but it is designed for managing Python packages. Here, you use `conda` to install `ipython`. Astute Pythonistas will notice that we have eschewed configuring any sort of virtual environment.

Define the Default Runtime Command

Your final instruction in the ipython Dockerfile is to give the image a default runtime command via the CMD[7] instruction. Each Dockerfile can contain a single CMD instruction, although this instruction is not required. Per the Dockerfile best practices,[8] the CMD instruction should be used to run the software contained by an image and should always be used in the form CMD ["executable", "param1", "param2"...]. In Listing 5-47, you add an instruction to your Dockerfile to execute ipython at runtime.

Listing 5-47. ipython/Dockerfile

```
FROM miniconda3
LABEL maintainer=@joshuacook
RUN conda update conda && \
    conda install --quiet --yes ipython && \
    conda clean -tipsy
CMD ["ipython"]
```

Build the ipython Image

In Listing 5-48, having defined the image using a Dockerfile, you build the image using the docker build command, again naming the image and providing the ipython directory as context.

Listing 5-48. Run a Docker Build

```
$ docker build -t ipython ipython
Sending build context to Docker daemon 2.048 kB
Step 1/4 : FROM jupyter-base
 ---> 014cc8d97486
Step 2/4 : LABEL maintainer @joshuacook
 ---> Using cache
 ---> 938a406795a2
Step 3/4 : RUN conda update conda &&     conda install --quiet --yes ipython
&&     conda clean -tipsy
 ---> Running in 64910e75091f
Fetching package metadata ........
Solving package specifications: .

Package plan for installation in environment /opt/conda:
```

---

[7]https://docs.docker.com/engine/reference/builder/#cmd
[8]https://docs.docker.com/engine/userguide/eng-image/
dockerfile_best-practices/#cmd

The following packages will be UPDATED:

```
    conda: 4.3.11-py36_0 --> 4.3.14-py36_0
```

```
...
 ---> f9a032f0a9a5.
Removing intermediate container 64910e75091f
Step 4/4 : CMD ipython
 ---> Running in 5839183d3d00
 ---> 98f0b133173e
Removing intermediate container 5839183d3d00
Successfully built 98f0b133173e
Successfully tagged ipython:latest
```

Commit the Changes to GitHub

In Listing 5-49, you commit this new image and push the changes to GitHub.

Listing 5-49. Commit the Changes and Push to GitHub

```
$ git add ipython/Dockerfile
$ git commit -m 'IPYTHON. Initial build'
$ git push
```

Run the ipython Image as a New Container

Finally, having developed your design and built your ipython image, let's run the image as a new container. In Listing 5-50, you run ipython as an interactive terminal containerized process. You terminate the process via Ctrl+D.

Listing 5-50. Run the ipython Image as a New Interactive Terminal Containerized Process

```
$ docker run -it ipython
Python 3.6.0 |Continuum Analytics, Inc.| (default, Dec 23 2016, 12:22:00)
Type "copyright", "credits" or "license" for more information.

IPython 5.3.0 -- An enhanced Interactive Python.
?         -> Introduction and overview of IPython's features.
%quickref -> Quick reference.
help      -> Python's own help system.
object?   -> Details about 'object', use 'object??' for extra details.

In [1]: import requests

In [2]: resp = requests.get('http://google.com')
```

```
In [3]: resp.status_code
Out[3]: 200

In [4]:
Do you really want to exit ([y]/n)?
```

Summary

In this chapter, I introduced the Dockerfile, the file type used to define an image in the Docker ecosystem. You explored several common instructions used in the definition of Dockerfiles. You also defined three images, the third built using the second as its base image. After reading this chapter, you should be familiar with the definition of new images using Docker best practices.

CHAPTER 6

■ ■ ■

Docker Hub

Equipped with tools for developing our own images, it quickly becomes important to be able to save and share the images we have written beyond our system. Docker Registries allow us to do just this. For your purposes, the public Docker Registry, Docker Hub, will be more than sufficient, though it is worth noting that other registries exist and that it is possible to create and host your own registry.

Overall, the Docker Registry consists of three technologies: the Index, the Registry, and the Repository. In this text, a cursory understanding of how these technologies interact is sufficient. The Index tracks meta-information associated with users, organizations, namespaces, and repositories. The Repository is similar to a git repository in that it tracks multiple versions of a project (that is, a Docker image). The Registry contains the images and the repository graph comprising a Repository.

Docker Hub

A Docker Registry is a server-side application that can be used to distribute Docker images. We are most interested in the free-to-use, public Docker Registry, Docker Hub.[1] Docker Hub is to Docker somewhat as GitHub is to Git. It allows us to use existing Docker repositories and allows us to build and host our own images. It will serve for the vast majority of your work as your sole Docker Registry from which you will discover the images you will use, manage the development of your images, and automate the build process as you move toward production-ready images.

Alternatives to Docker Hub

Two of the most popular alternative public registries are Quay.io[2] and the Google Container Registry.[3] Quay.io is a public and private registry service run by CoreOS. CoreOS is the developer of the primary container engine alternative to Docker, rkt.[4] In the past year, however, CoreOS and Docker seem to be working more as allies than competitors and Quay.io can serve as a registry for both Docker and rkt-defined images. Quay has tools for the maintenance of images by organizations and teams plus workflow automation tools, and it is an excellent alternative to Docker Hub.

---

[1] http://hub.docker.com/
[2] http://quay.io/
[3] http://gcr.io/
[4] https://coreos.com/rkt

© Joshua Cook 2017
J. Cook, *Docker for Data Science*, DOI 10.1007/978-1-4842-3012-1_6

Google was an early organizational adopter of containerization technology and is the origin of one of the community's most beloved tools, the orchestration tool Kubernetes. It is no surprise that Google has its own container registry, the Google Container Registry. For our purposes, that is, for the purposes of data science, we might most be interested in using the GCR as the home of the in-house images for Google's machine intelligence library, Tensorflow.[5]

Docker ID and Namespaces

In order to leverage the services offered by Docker Hub, you will create a Docker ID. A new Docker ID can be created via the Docker Cloud sign-up page.[6] Creation of a Docker ID will require email verification.

Once created, your Docker ID also becomes your main namespace you will use for all of your images hosted on Docker Hub. My Docker ID is joshuacook. Were I to push the gsl image we created in Chapter 4 to Docker Hub, this image would be available at the namespace/tag combination of joshuacook/gsl. More generally, the namespace/tag of an image hosted on Docker Hub will appear as in Listing 6-1.

Listing 6-1. General Namespace/Tag for an Image on Docker Hub

```
<namespace>/<repository_name>:<tag>
```

Were you to use an image not hosted on Docker Hub, such as the Tensorflow GPU image as in Listing 6-2, you would need to specify the full URI including the registry address, following the pattern outlined in Listing 6-3.

Listing 6-2. Full URI for the Latest Tensorflow GPU Image

```
gcr.io/tensorflow/tensorflow:latest-gpu
```

Listing 6-3. General URI for a Registry-Hosted Image

```
<registry_address>/<namespace>/<repository_name>:<tag>
```

Image Repositories

A repository on Docker Hub is a collection of tagged, built Docker images sharing the same purpose. Visiting my user page, you can see all of the Docker repositories I am currently maintaining (Figure 6-1). Each repository contains one or more images that have been previously defined and built, either locally or via the Docker Hub Automated Build process.

---

[5]www.tensorflow.org/
[6]https://cloud.docker.com/

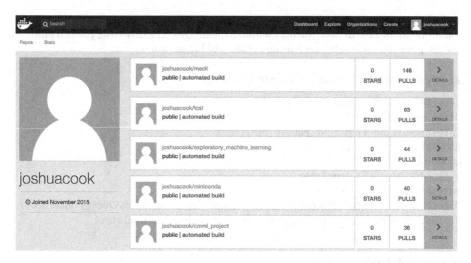

Figure 6-1. My Docker Hub user profile

Search for Existing Repositories

As previously noted, the Docker Hub registry is the default registry used by the Docker CLI. As such, it is a trivial process to search the Docker Hub for relevant images using the Docker CLI using the docker search command. A keyword search via the docker search command checks the keyword against image name, users, and organizations, as well as an image description. In Listing 6-4, you search Docker Hub for miniconda images.

■ **Note** Images are returned in descending order based upon their number of stars.

Listing 6-4. Search Docker Hub for miniconda Images

```
$ docker search miniconda
NAME                         DESCRIPTION                     STARS
OFFICIAL    AUTOMATED
continuumio/miniconda        Powerful and flexible package ... 38
[OK]
alaindomissy/docker-miniconda docker-miniconda                2
[OK]
yamitzky/miniconda-neologd   Dockernized mecab-ipadic-NEolo... 2
[OK]
showOk/alpine-miniconda      An alpine based image with min... 1
[OK]
kentwait/miniconda-mpi       Docker container for developin... 1
[OK]
pottava/miniconda            Miniconda images based on Alpi... 1
[OK]
```

The search returns results both where the search term miniconda appears in the name and in the description. Having identified the image you wish to use, you can pull the image as you did in Chapter 3 using docker pull (Listing 6-5).

Listing 6-5. Pull the continuumio/miniconda Image

```
$ docker pull continuumio/miniconda
Using default tag: latest
latest: Pulling from continuumio/miniconda
8ad8b3f87b37: Pull complete
090d0f0e845b: Pull complete
3cc1bbd57a94: Pull complete
bd7b36ac12a3: Pull complete
Digest: sha256:f7e0a8a86a6d194e748c5884f53ddbbde33b08a666bed5370e453f35bbc3ec57
Status: Downloaded newer image for continuumio/miniconda:latest
```

Tagged Images

There is an active,[7] well-known, and friendly split in the Python community between versions 2 and 3. The implications of this are that any Python technology must be capable of being run against two major versions of Python. On your personal computer, this is a significant task requiring package managers supporting both Python 2 and 3 and a virtual environment system such as virtualenv or that provided by conda. With Docker, this task is trivially managed with a Docker image tag.

Image tags define variations in the definition of an image under a single namespace and repository combination. Tags have no semantic meaning, nor does your Docker Id or the name of your repository. They serve solely to distinguish between subtle changes made to images.

An image can be given a tag in any of three ways:

1. An image can be tagged at build time simply by appending :<tag> to the end of the name given to the image (Listing 6-6). Using this method, a tagged image will be associated with a specific Dockerfile.

2. An image can be tagged afterward using the docker tag command (Listing 6-7). Using this method, a tagged image will be associated with a specific Dockerfile.

3. Ephemeral changes made to a container can be persisted as a new image using the docker commit command (Listing 6-8). Using this method, no Dockerfile will exist describing the ephemeral changes made to the image.

---

[7] The latest version of IPython, however, does not support Python 2 (http://ipython.readthedocs.io/en/stable/).

Listing 6-6. Tag an Image During a Build

```
$ docker build -t <namespace>/<repository_name>:<tag>
```

Listing 6-7. Retag an Existing Local Image

```
$ docker tag <existing_image> <namespace>/<repository_name>:<tag>
```

Listing 6-8. Commit Changes to a Container as a Tagged Image

```
$ docker commit <existing_container> <namespace>/<repository_name>:<tag>
```

You will revisit this last as a best practice in maintaining semipersistent changes to images in the next chapter.

Tags on the Python Image

The python image can be found at the official repository page[8] for Python. Visiting the Tags tab, you can see specific information about the dozens of tags associated with the Python repository. The Python repository uses image tags to not only manage the Python 2 and Python 3 split, but to manage four different versions of Python 3. From the official repository page, you can link to the Dockerfiles used to define the individually tagged images in the Python repository (see Figure 6-2).

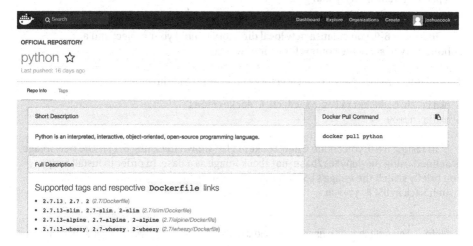

Figure 6-2. *The Python Official Repository page*

---

[8]https://hub.docker.com/_/python/

Official Repositories

Docker maintains a set of curated Docker images for the major open-source technologies. These official repositories are designed using established best practices in writing Dockerfiles (and more likely than not are maintained by tianon[9]). You saw your first official repository in the Official Repository for Python. With the exception of the jupyter image (which you shall treat as though it is an Official Repository), you will nearly always use Official Repositories as your base image.

Pushing to Docker Hub

To demonstrate the process of pushing to Docker Hub, you will create a new image, numpy-notebook. This image will use the jupyter/base-notebook image as its base and add numpy. Recall that the jupyter/base-notebook image includes only Python 3. You will only add numpy for Python 3.

■ **Note**   In Chapter 5, you configured your system for local development by setting up a project to be tracked via git and GitHub. This is a best practice. You will not go through this practice for the creation of an image in this chapter. This is not because it is not a good practice, but rather because I wish to emphasize that working with Docker Hub is independent from working with git.

In Listing 6-9, you create a new local directory to hold your project and a subdirectory to serve the context for your new image.

Listing 6-9. Create a New Local Directory and Context Subdirectory

```
$ mkdir ch_6_dockerfiles && cd ch_6_dockerfiles
$ mkdir numpy
```

In Listing 6-10, you create a new Dockerfile (shown in Listing 6-11). The new image is defined using the jupyter/base-notebook image as a base. In order to install numpy you briefly switch the image's USER to root. You use conda to install numpy and then switch back to USER jovyan.

■ **Note**   You switch to the user root to install the libraries and switch back to user jovyan upon completion. This is considered a best practice and ensures that you do not run the notebook server with too much system privilege. More on this in Chapter 7.

---

[9]https://github.com/tianon

Listing 6-10. Create a New Dockerfile

```
$ vi numpy/Dockerfile
```

Listing 6-11. The numpy Dockerfile

```
FROM jupyter/base-notebook
USER root
RUN conda install --yes numpy
USER jovyan
```

In Listing 6-12, you build the numpy image using the docker build command. Note that the term numpy shows up twice in the command. The first (-t numpy) refers to the tag or name that you are giving to the image. The second, the last word in the command, refers to the relatively referenced build context, in this case the subdirectory named numpy.

Listing 6-12. Build the numpy Image

```
$ docker build -t numpy numpy
Sending build context to Docker daemon  2.048kB
Step 1/4 : FROM jupyter/base-notebook
...
Step 2/4 : USER root
...
Step 3/4 : RUN conda install numpy
...
Step 4/4 : USER jovyan
...
Successfully built 2570ccf8069f
Successfully tagged numpy:latest
```

In Listing 6-13, you run the ipython REPL using the numpy image you just built as a means of testing your installation. You use Ctrl+D to exit the containerized process after completing a trivial calculation.

Listing 6-13. Run ipython from the numpy Image

```
$ docker run -it numpy ipython
Python 3.6.1 | packaged by conda-forge | (default, May 23 2017, 14:16:20)
Type 'copyright', 'credits' or 'license' for more information
IPython 6.1.0 -- An enhanced Interactive Python. Type '?' for help.

In [1]: import numpy as np

In [2]: u = np.array((1,2))

In [3]: np.linalg.norm(u)
Out[3]: 2.2360679774997898

In [4]:
Do you really want to exit ([y]/n)?
```

Create a New Repository

Creating a new repository can be done via a Docker Hub user profile (see Figure 6-3). You will be prompted to give the repository a name, a short description, and a full description. You will also be asked whether the repository should be public or private. By default, new repositories will be public.

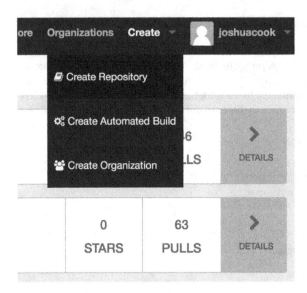

Figure 6-3. *Create a new repository*

Again, there is no semantic meaning to the name given to a repository. With regard to the description, you should recall that keywords entered in the description were found by the docker search function.

■ **Note**   You could also choose to create the repository in any organization where you have the proper privileges.

You create a new repository called numpy with the short description "Numerical Python" and the full description "Built on jupyter/base-notebook." The new repository is shown in Figure 6-4.

Figure 6-4. *The new joshuacook/numpy repository*

Push an Image

You might wish to push the work you did on your numpy image. Local images created in the manner discussed in Chapter 5 can be pushed to Docker Hub with little fuss, provided they are named in the <namespace>/<repository_name>:<tag> pattern. In order to push your work, you first revisit the state of the images on your system via the docker images command (Listing 6-14), especially to examine the numpy image you just created.

Listing 6-14. Display Local Images

```
$ docker images
REPOSITORY          TAG        IMAGE ID         CREATED          SIZE
numpy               latest     2570ccf8069f     33 minutes ago   925MB
jupyter/            latest     161472bc6c75     2 weeks ago      657MB
base-notebook
debian              latest     47af6ca8a14a     2 weeks ago      125.1 MB
miniconda3          latest     5865a6cfa8c2     2 weeks ago      1.64 GB
```

───

■ **Note**   None of your images have a namespace or a tag.

───

Again, there are three ways to give a tag to an image: 1) at build-time using `docker build` 2) after build-time using `docker tag`, and 3) by committing changes made to a container as an image using `docker commit`. In Listing 6-15, you give the numpy image a tag and a namespace using the `docker tag` command.

Listing 6-15. Tag an Image

```
$ docker tag numpy joshuacook/numpy:1.13.0
```

In Listing 6-16, you once more display your local images.

Listing 6-16. Display Local Images

```
$ docker images
REPOSITORY       TAG         IMAGE ID        CREATED          SIZE
numpy            latest      2570ccf8069f    34 minutes ago   925MB
joshuacook/      1.13.0      2570ccf8069f    34 minutes ago   925MB
numpy
jupyter/         latest      161472bc6c75    2 weeks ago      657MB
base-notebook
debian           latest      47af6ca8a14a    2 weeks ago      125.1MB
miniconda3       latest      5865a6cfa8c2    2 weeks ago      1.64GB
```

You now have an image with the namespace/repository combination of `joshuacook/numpy` and the tag of `1.13.0`. Note that it has an identical image id with the numpy image. Docker is not storing two identical images, but rather maintaining two references to the same image.

In order to push to Docker Hub, you must log in via the Docker CLI using `docker login` (Listing 6-17).

Listing 6-17. Log in to Docker Hub

```
$ docker login
Login with your Docker ID to push and pull images from Docker Hub. If you
don't have a Docker ID, head over to https://hub.docker.com to create one.
Username (joshuacook):
Password:
Login Succeeded
```

Once you have properly tagged your image and logged in, you use the `docker push` command (Listing 6-18).

Listing 6-18. Push the ipython Image to Docker Hub

```
$ docker push joshuacook/numpy:1.13.0
The push refers to a repository [docker.io/joshuacook/numpy]
2bcb5be18c74: Pushed
72c56d765cf2: Mounted from jupyter/base-notebook
317e6c337ef3: Mounted from jupyter/base-notebook
1a2fceb0b4da: Mounted from jupyter/base-notebook
c65670cc3813: Mounted from jupyter/base-notebook
52b6625b711f: Mounted from jupyter/base-notebook
93329dd321d3: Mounted from jupyter/base-notebook
7c10c4ddeab0: Mounted from jupyter/base-notebook
290b555a5673: Mounted from jupyter/base-notebook
6b538a724de5: Mounted from jupyter/base-notebook
f02f8903fe33: Mounted from jupyter/base-notebook
406306ca7a80: Mounted from jupyter/base-notebook
cff5883220e6: Mounted from jupyter/base-notebook
d17d48b2382a: Mounted from jupyter/base-notebook
1.13.0: digest: sha256:ccd8f21923c7538ae7a4d0606e203dce072d601494e570cf4c3d1
d08ca7a84e2 size: 3246
```

Pull the Image from Docker Hub

You can verify the success of your push by pulling the image from Docker Hub. Of course, if you have the image locally, the locally cached image will be used when the image is requested. In order to test the success, you first inspect the contents of your local image using the docker inspect command in order to use this for verification (Listing 6-19). You are interested in the "RootFS" key in the JSON returned by the inspect command, as it provides a sha256 description[10] description of each layer associated with the image.

Listing 6-19. Inspect the ipython Image

```
$ docker inspect joshuacook/numpy:1.13.0
...
        "RootFS": {
            "Type": "layers",
            "Layers": [
                "sha256:d17d48b2382adda1fd94284c51d725f0226bf20b07f4d29ce09
                596788bed7e8e",
                "sha256:cff5883220e61c711e6345366431e2eb28d8b408ae02c21c135
                6797932379f7f",
                "sha256:406306ca7a8025fd3430c01289c18e5ed18f0a144e7b6f1bc59
                2ad38817f52a3",
                "sha256:f02f8903fe334cbe7184c6d57fc08e6b5d26b607fce64c090c0
                79d2a996f14a3",
```

---

[10]https://en.wikipedia.org/wiki/SHA-2

```
      "sha256:6b538a724de5d6ecdbfa4583dc34a228a46c3ee0c4d309a481e
      9dfad675de380",
      "sha256:290b555a56733ef2f2a005e6c7a3c38d894674239aed4777290
      92687c414015d",
      "sha256:7c10c4ddeab02a978638181a43ac67d43036fc6bf67e9888544
      debbd63aa11b3",
      "sha256:93329dd321d38f8394e015b422cf3680be1de5568f7248a3b63
      5df329b2fe47b",
      "sha256:52b6625b711fbf05039b819e2d13161f5c36c1909ad61779efe
      dae05a5fdc51c",
      "sha256:c65670cc38137214111c9e1587cb200e32e74de13fc2957752d
      6354f75da6278",
      "sha256:1a2fceb0b4daff636aa021a223b46c308a94e52f084c9feea39
      5b68f672be6cb",
      "sha256:317e6c337ef3c57618c38257236cc02e254f2f8d499249fbc04
      0208f25c360d9",
      "sha256:72c56d765cf2ae7ce7626b5a35bf0eba94f8c49b1a8d894b999
      949846b2ded71",
      "sha256:2bcb5be18c742a706f4667ce18b59e45d777e01d2423aac3c03
      5c0d2831e34fc"
   ]
 }
...
```

You will use these layer descriptions to verify against the image once you have pulled it from Docker Hub.

Next, you attempt to remove the joshuacook/ipython:3.6 image from your local cache (Listing 6-20).

Listing 6-20. Remove the Image from Local Cache

```
$ docker rmi numpy
Untagged: numpy:latest
$ docker rmi joshuacook/numpy:1.13.0
Error response from daemon: conflict: unable to remove repository reference
"joshuacook/numpy:1.13.0" (must force) - container 817ba39439d7 is using its
referenced image 2570ccf8069f
```

In doing so, you see an error. This error signifies that a stopped container is using the image. You must first remove the stopped container in order to remove the image. You first display stopped containers via the -a flag (display all) and docker ps (Listing 6-21).

Listing 6-21. Display All Containers

```
$ docker ps -a
CONTAINER ID
IMAGE          COMMAND          CREATED    STATUS             NAMES
817ba39439d7   "tini -- ipyth"  29 min...  Exited (0) 28 min...  gifted_clarke
2570ccf8069f
fd66407a5358   "tini -- ipyth"  38 min...  Exited (0) 36 min...  peaceful_
2570ccf8069f                                                     ardinghelli
...
```

Two stopped containers reference image f9a032f0a9a5, your local ipython image. You must remove all of them (Listing 6-22).

Listing 6-22. Remove All Stopped ipython Containers

```
$ docker rm 817ba39439d7 fd66407a5358
fd66407a5358
fd66407a5358
```

Now you are able to remove the locally cached image (Listing 5-23).

Listing 6-23. Remove the Image from Local Cache

```
$ docker rmi joshuacook/ipython:3.6
Untagged: joshuacook/numpy:1.13.0
Untagged: joshuacook/numpy@sha256:ccd8f21923c7538ae7a4d0606e203dce072d601494
e570cf4c3d1d08ca7a84e2
Deleted: sha256:2570ccf8069f2333fc3c52cdeca9890dd4cb6b7a27ee7752aa97d8a2bc4
e1bf6
Deleted: sha256:b66f2cee6b59bb573462448ee4766890737e74523fa3d14b99a418c57a1
e67f1
Deleted: sha256:fe5d8ad29a14a9f6405de89aa47ef8c511965aff9d8b20cbeeffe34ee6f
e6f19
Deleted: sha256:eddb59d7cee362460c5d62be665d004d008f724a3e65ad44ad1129cfb86
f3f61
```

You now pull the image from Docker Hub to verify its contents (Listing 6-24).

Listing 6-24. Pull the ipython Image from Docker Hub

```
$ docker pull joshuacook/numpy:1.13.0
1.13.0: Pulling from joshuacook/numpy
693502eb7dfb: Already exists
490c0d36e714: Already exists
b47c251cda4e: Already exists
5f06af7aed8b: Already exists
6486d270a020: Already exists
825ae89ffbbc: Already exists
```

```
0eb855700e1f: Already exists
3ea165122423: Already exists
57f4c53afea9: Already exists
960ee91f3ec0: Already exists
d685ecb69227: Already exists
86a69e035999: Already exists
166ce3ece426: Already exists
9c7191cb9c0e: Pull complete
Digest: sha256:ccd8f21923c7538ae7a4d0606e203dce072d601494e570cf4c3d1d08ca7a
84e2
Status: Downloaded newer image for joshuacook/numpy:1.13.0
```

Once more you run the docker inspect command (Listing 6-25).

Listing 6-25. Inspect the numpy Image After a Successful Pull

```
$ docker inspect joshuacook/numpy:1.13.0
...
        "RootFS": {
            "Type": "layers",
            "Layers": [
                "sha256:d17d48b2382adda1fd94284c51d725f0226bf20b07f4d29ce09
                596788bed7e8e",
                "sha256:cff5883220e61c711e6345366431e2eb28d8b408ae02c21c135
                6797932379f7f",
                "sha256:406306ca7a8025fd3430c01289c18e5ed18f0a144e7b6f1bc59
                2ad38817f52a3",
                "sha256:f02f8903fe334cbe7184c6d57fc08e6b5d26b607fce64c090c0
                79d2a996f14a3",
                "sha256:6b538a724de5d6ecdbfa4583dc34a228a46c3ee0c4d309a481e
                9dfad675de380",
                "sha256:290b555a56733ef2f2a005e6c7a3c38d894674239aed4777290
                92687c414015d",
                "sha256:7c10c4ddeab02a978638181a43ac67d43036fc6bf67e9888544
                debbd63aa11b3",
                "sha256:93329dd321d38f8394e015b422cf3680be1de5568f7248a3b63
                5df329b2fe47b",
                "sha256:52b6625b711fbf05039b819e2d13161f5c36c1909ad61779efe
                dae05a5fdc51c",
                "sha256:c65670cc38137214111c9e1587cb200e32e74de13fc2957752d
                6354f75da6278",
                "sha256:1a2fceb0b4daff636aa021a223b46c308a94e52f084c9feea39
                5b68f672be6cb",
                "sha256:317e6c337ef3c57618c38257236cc02e254f2f8d499249fbc04
                0208f25c360d9",
                "sha256:72c56d765cf2ae7ce7626b5a35bf0eba94f8c49b1a8d894b999
                949846b2ded71",
```

```
        "sha256:2bcb5be18c742a706f4667ce18b59e45d777e01d2423aac3c03
        5c0d2831e34fc"
    ]
}
```
...

Note that the sha256 description of the images layers after the pull matches the
description of the layers prior to the pull. It is of note that the pull proves to be rather trivial,
only needing to fetch a single layer from Docker Hub. This is because most of the layers
associated with your numpy image exist as the jupyter/base-notebook image. You used the
jupyter/base-notebook image as the base upon which you built your ipython image! This
fact can be verified by inspecting the jupyter/base-notebook image (Listing 6-26).

Listing 6-26. Inspect the jupyter/base-notebook Image

```
$ docker inspect jupyter/base-notebook
...
        "RootFS": {
            "Type": "layers",
            "Layers": [
                "sha256:d17d48b2382adda1fd94284c51d725f0226bf20b07f4d29ce09
                596788bed7e8e",
                "sha256:cff5883220e61c711e6345366431e2eb28d8b408ae02c21c135
                6797932379f7f",
                "sha256:406306ca7a8025fd3430c01289c18e5ed18f0a144e7b6f1bc59
                2ad38817f52a3",
                "sha256:f02f8903fe334cbe7184c6d57fc08e6b5d26b607fce64c090c0
                79d2a996f14a3",
                "sha256:6b538a724de5d6ecdbfa4583dc34a228a46c3ee0c4d309a481e
                9dfad675de380",
                "sha256:290b555a56733ef2f2a005e6c7a3c38d894674239aed4777290
                92687c414015d",
                "sha256:7c10c4ddeab02a978638181a43ac67d43036fc6bf67e9888544
                debbd63aa11b3",
                "sha256:93329dd321d38f8394e015b422cf3680be1de5568f7248a3b63
                5df329b2fe47b",
                "sha256:52b6625b711fbf05039b819e2d13161f5c36c1909ad61779efe
                dae05a5fdc51c",
                "sha256:c65670cc38137214111c9e1587cb200e32e74de13fc2957752d
                6354f75da6278",
                "sha256:1a2fceb0b4daff636aa021a223b46c308a94e52f084c9feea39
                5b68f672be6cb",
                "sha256:317e6c337ef3c57618c38257236cc02e254f2f8d499249fbc04
                0208f25c360d9",
                "sha256:72c56d765cf2ae7ce7626b5a35bf0eba94f8c49b1a8d894b999
                949846b2ded71"
            ]
        }
...
```

■ **Note**   The thirteen layers that comprise the `jupyter/base-notebook` image are exactly the thirteen layers of the `numpy` image.

Tagged Image on Docker Hub

Finally, let's visit the Tags tab on the Docker Hub page associated with your numpy repository (Figure 6-5). Here, you can see that that tagged image you pushed is indeed available on Docker Hub.

Figure 6-5. *The joshuacook/numpy Docker Hub page*

Summary

In this chapter, you looked at using Docker Hub as a cloud-based store for the images you have built. You learned about multiple versions of images, looking at how tags are used to track multiple versions of the Python community image. You created a new image for using the numpy library built using the `jupyter/base-notebook` image as a base, tagged it with the current version of numpy, and pushed it to your Docker Hub account. After this chapter, I hope that you are familiar with the process of creating a new image from a `Dockerfile` and know how to push an image that you create to Docker Hub.

CHAPTER 7

■ ■ ■

The Opinionated Jupyter Stacks

The Jupyter Notebook is based on a set of open standards for interactive computing.

—Project Jupyter[1]

Project Jupyter developed out of an academic environment and out of the project came not simply a groundbreaking application, but also a well-defined set of protocols for the interactive computing paradigm. With regard to the notebook, these protocols include the notebook document format, an interactive computing protocol, and the kernel. These and the other protocols defining the Jupyter ecosystem are maintained by an openly-governed steering council.[2] Beyond the protocols defining the interactive computing via Jupyter and maintaining the numerous projects that comprise the ecosystem, Project Jupyter also maintains a GitHub repository containing numerous well-defined Jupyter Docker images: the Opinionated Jupyter Stacks.[3]

It is a trivial endeavor to launch a Jupyter Notebook server within a Docker container. In Listing 7-1, you do this using the -P (publish all) flag, which publishes all exposed ports to random ports on the host.

Listing 7-1. Launch a Jupyter `scipy-notebook` Server

```
$ docker run -d -P jupyter/scipy-notebook
72da8ca9ac9d4c694477350c500e9d793769788a26148a144e6c29448b5b4840
```

To use this container, you will need to know to which port on your host system port 8888 in the container has been published (Listing 7-2) and will need to obtain the notebook server's security token (Listing 7-3). In Listing 7-2, you use the `docker port` command to obtain the port mappings for your container using the first four characters, `72da`, of the `container id` returned when you started the container, `72da8ca9ac9d4....`

---

[1]http://jupyter.org/
[2]http://jupyter.org/about.html
[3]https://github.com/jupyter/docker-stacks

© Joshua Cook 2017
J. Cook, *Docker for Data Science*, DOI 10.1007/978-1-4842-3012-1_7

■ **Note**   A Docker container can be referenced by as few characters are required to establish a unique container reference. The container, 72da8ca9ac9d4c694477350c500e9d793769788a26148a144e6c29448b5b4840, could be referenced by 7, if no other container ids begin with 7.

Listing 7-2. Obtain a Port Mapping for the Container

```
$ docker port 72da
8888/tcp -> 0.0.0.0:32769
```

In Listing 7-3, you use the docker exec command to send a command to your Jupyter Notebook server within the docker container where it is being run. You again use the first four characters of the container id to reference the container and then send the container the command jupyter notebook list. This has the effect of opening a shell to the container, running this command in the shell, displaying the result, and closing the shell.

Listing 7-3. Obtain a Security Token for Running Jupyter Notebook Server

```
$ docker exec 72da jupyter notebook list
Currently running servers:
http://localhost:8888/?token=61d6ead40b05daea402d9843ad7932bc937
da41841575765 :: /home/jovyan
```

Bear in mind that the URL provided here refers to the URL where the notebook server can be reached within the Docker container. The actual location would be to use the IP of your docker machine (localhost or the IP of your AWS instance or the virtual machine on which you are running Docker) and the port you obtained in the previous command. Listing 7-4 shows this URL if using Docker for Linux, Docker for Mac, or Docker for Windows. Listing 7-5 shows the most likely URL if using Docker Toolbox. (Refer to Listing 1-10 in Chapter 1 to review accessing your Jupyter system on AWS.)

Listing 7-4. The Current Jupyter URL If Using Docker for Linux/Mac/Windows

```
http://localhost:32769/?token=61d6ead40b05daea402d9843ad7932bc937
da41841575765
```

Listing 7-5. The Likely Current Jupyter URL If Using Docker for Toolbox

```
http://192.168.99.100:32769/?token=61d6ead40b05daea402d9843ad7932bc937
da41841575765
```

High-Level Overview

The Project Jupyter Docker stacks provide eight notebook images beginning with a base-notebook image upon which all subsequent images depend:

1. base-notebook

2. minimal-notebook

3. scipy-notebook

4. r-notebook

5. tensorflow-notebook

6. datascience-notebook

7. pyspark-notebook

8. all-spark-notebook

Each image has an eponymous folder in the GitHub jupyter/docker-stacks project, as well as an equivalent repository hosted under the jupyter user namespace on Docker Hub. Figure 7-1 shows the dependencies of these images.

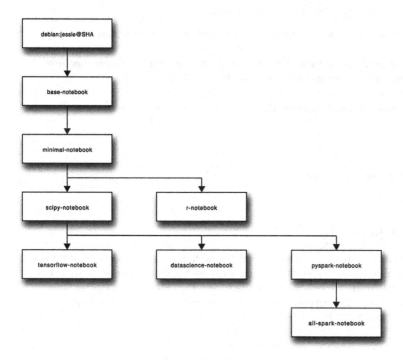

Figure 7-1. *Jupyter Docker stack dependency graph*

■ **Note**   The base-notebook uses Debian "jesse"[4] and thus provides the container operating system for every notebook image.

---

jupyter/base-notebook

The jupyter/base-notebook image defines a minimal Jupyter Notebook server. It is solely provisioned with Miniconda for Python 3 and does not come provisioned with Python 2 nor any scientific computing packages (from numpy on up). It does define the essential patterns that will be used to define Jupyter images through the entire stack: the use of the init binary tini, the addition of the unprivileged user jovyan,[5] and the inclusion of various startup scripts that will be used to run the server.

You will rarely use the base-notebook image in practice, but will take the development patterns written here as best practices. Any changes you might make to an image downstream will need to be made considering these practices. You have done so already when you mounted a volume of local files to your server being run in a container in Chapter 2. You mounted these files from your working directory to the container directory /home/jovyan using the flag -v `pwd`:/home/jovyan, as in Listing 7-6.

Listing 7-6. Attach a Host Directory to a Container

```
$ docker run -d -v `pwd`:/home/jovyan -P jupyter/scipy-notebook
```

The use of the /home/jovyan directory as root notebook directory is defined in the Dockerfile defining the base-notebook image (Listing 7-7).

Listing 7-7. base-notebook Dockerfile

```
...
ENV NB_USER jovyan
...
WORKDIR /home/$NB_USER/work
...
```

Notebook Security

It is not uncommon to run a notebook server as a publicly accessible server over the open web. Furthermore, a notebook server is capable of running arbitrary code. As such, it is a best practice[6] to restrict access to a notebook server. The Opinionated Jupyter Stacks define a series of security best practices, largely defined in the Dockerfile for the base-notebook image.

---

[4]www.debian.org/releases/jessie/
[5]Jovyan means "related to Jupiter" (Jupyter).
[6]http://jupyter-notebook.readthedocs.io/en/latest/security.html

By default, a notebook server running in a container requires a randomly generated security token to be passed as a query parameter when accessed through the browser. You saw earlier how this token can be obtained by sending the jupyter notebook list command to a running container via docker exec (Listing 7-3). The base-notebook image defines the start-notebook.sh script used as the default command passed to tini at runtime. This script can be used to define an alternative authentication method, although this is not recommended.

In an installation in which you can guarantee security (or in which you are comfortable being somewhat less restrictive, such as a temporary AWS instance), you may wish to grant sudo access to the jovyan user. This can be done by passing the environment variable GRANT_SUDO=yes. Additionally, the container must be run by the root user. Listing 7-8 shows a complete command for running a notebook server and granting password-less sudo to the jovyan user.

Listing 7-8. Run a Notebook Server and Grant Password-less sudo to the jovyan User

```
$ docker run -d -e GRANT_SUDO=yes --user root jupyter/scipy-notebook
a811689f2e09737e2c9686849320a424889b2ac9eeb57e0f3df2940edc600628
```

Granting sudo access to the jovyan user is useful while working in an exploratory capacity. You may wish to quickly install Linux binaries that are not part of your image without going through a full image build cycle. With sudo access granted, this is possible.

If you connect to the running container via docker exec, you can see that tini has been launched by the root user (Listing 7-9). The GRANT_SUDO flag has the effect of launching the jupyter notebook process, owned by jovyan, via su jupyter notebook.

Listing 7-9. Connect to a Running Container as root and View the Running Processes

```
$ docker exec -it a811 ps aux
USER   PID  %CPU %MEM VSZ    RSS   TTY STAT START TIME COMMAND
root   1    1.0  0.0  4224   668   ?   Ss   23:41 0:00 tini -- start
                                                       -notebook.sh
root   6    0.4  0.1  46360  3084  ?   S    23:41 0:00 su jovyan -c
                                                       env ...
jovyan 9    33.2 2.4  183488 50612 ?   Ss   23:41 0:01 /opt/conda/
                                                       bin/python ...
root   13   0.0  0.1  19100  2556  ?   Rs   23:41 0:00 ps aux
```

A notebook server running without this flag (Listing 7-10) will appear as in Listing 7-11. Note that tini has been launched by jovyan and that no su command has been run to grant sudo to jovyan.

Listing 7-10. Run a Notebook Server

```
$ docker run -d jupyter/base-notebook
e900cbb66babb23f8b7764506482e32e300b8ad351e4ea15a0260266ca517738
```

Listing 7-11. Connect to a Running Container as jovyan and View the Running Processes

```
$ docker exec e900 ps aux
USER     PID  %CPU  %MEM  VSZ     RSS    TTY STAT  START TIME  COMMAND
jovyan   1    0.0   0.0   4224    656    ?   Ss    23:35 0:00  tini -- start-
                                                                notebook.sh
jovyan   5    1.1   2.4   183404  50576  ?   S     23:35 0:01  /opt/conda/bin/
                                                                python ...
jovyan   2    0.0   0.1   19100   2436   ?   Rs    23:37 0:00  ps aux
```

The Default Environment

The base-notebook Dockerfile installs miniconda3 and the latest stable version of Python 3. It also defines the default conda environment, root, which will be used throughout to manage your Python 3 installation. You can see the miniconda version and the name of your Python 3 environment by running an interactive bash shell to a base-notebook container (Listing 7-12).

■ **Note** ipython, python, pip, easy_install, and conda are all available in this environment and reference the conda/python versions installed in the base-notebook Dockerfile.

Listing 7-12. Identify the Python 3 conda Environment.

```
$ docker run -it --rm jupyter/base-notebook bash
jovyan@c9fb2312c5b8:~$ conda info -a
Current conda install:

               platform : linux-64
          conda version : 4.2.12
       conda is private : False
      conda-env version : 4.2.12
    conda-build version : not installed
         python version : 3.5.2.final.0

...

# conda environments:
#
root                  *  /opt/conda
```

It is worth noting that you could have run the same command in an already running container, as in Listing 7-13. Listing 7-13 has the effect of running the command via a shell (as opposed to a bash shell) and will close the shell when the command has completed execution. Here, wizardly_hawking refers to the randomly generated name for a running container.

Listing 7-13. Alternatively Identify the Python 3 conda Environment

```
$ docker exec wizardly_hawking conda info -a
Current conda install:
```

...

Here you can see that you are running Python 3.5.2 and have a single Python environment root. You can verify that the root environment is running 3.5.2 by sourcing the root environment and checking the Python version (Listing 7-14). The source activate command has the effect of configuring the global python to use the python binary available at /opt/conda/bin/python.

Listing 7-14. Check the Python Version for root Environment

```
jovyan@ c9fb2312c5b8:~$ source activate root
(root) jovyan@ c9fb2312c5b8:~$ python --version
Python 3.5.2 :: Continuum Analytics, Inc.
(root) jovyan@ c9fb2312c5b8:~$ which python
/opt/conda/bin/python
```

Managing Python Versions

With the release of IPython 6 and the drop of support for Python 2, in this author's opinion there is no need to explicitly work with environments within a Docker image. It is perfectly reasonable to have one Docker image dedicated to running the latest stable version of Python, another image dedicated to running a legacy version of Python for an application that is useful but not worth updating, and still another image dedicated to running the latest alpha build of Python on all the same system. This is, in fact, a perfect use case for Docker. The dependencies of each implementation are completely isolated from each other when running in their own containers. For the time being, however, as the community continues to make the measured migration from Python 2 to Python 3, we will need to continue to explicitly manage environments running within our containers, for the primary purpose of maintaining simultaneous py2 and py3 environments. To do this, we will use conda's native capacity for managing environments.[7]

---

[7]https://conda.io/docs/using/envs.html

It is worth noting, before digging into the details of environment implementation, that switching environments in a running notebook or while creating a new notebook is a trivial task. We are investigating environments for the purposes of understanding the best way to add new libraries that are not installed by default. Figure 7-2 shows how new notebooks using different Python kernels can be created. Figure 7-3 shows how the kernel can be switched for a running notebook.

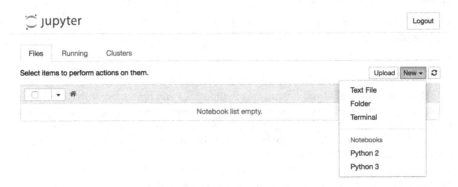

Figure 7-2. *Select a kernel when creating a new notebook*

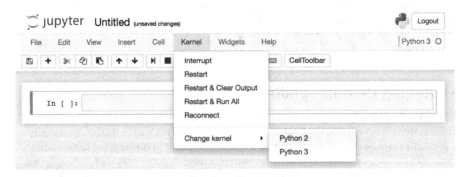

Figure 7-3. *Change the kernel in a running notebook*

Naively Extending a Jupyter Image using a `Dockerfile`

For the majority of the work you will do, you will take an image from the Jupyter Docker stacks and add libraries specific to your work as needed using the build process described in Chapter 4. Suppose, for example, that you wish to develop a semantic analysis project on Twitter data using the Python machine learning library `scikit-learn` and storing the data in a MongoDB database. `scikit-learn` is included by default with the `jupyter/scipy-notebook` image, but the python mongo interface library, pymongo, is not. A naive approach to building this image might use the `Dockerfile` described in Listing 7-15.

Listing 7-15. Dockerfile Extending the jupyter/scipy-notebook for Semantic Analysis

```
FROM jupyter/scipy-notebook
RUN pip install pymongo
```

You could then build the image using the docker build command (Listing 7-16) and run a container defined by this image using docker run (Listing 7-17).

Listing 7-16. Build the semantic_analysis Image

```
$ docker build -t semantic_analysis .
```

Listing 7-17. Run the semantic_analysis Container

```
$ docker run -d -P semantic_analysis
67cf8215ed6ad75c3709455bcdfffb72fb92df2abae3627962ba63677e06c45a
$ docker port 67cf
8888/tcp -> 0.0.0.0:32776
$ docker exec 67cf jupyter notebook list
Currently running servers:
http://localhost:8888/?token=d57ccd98004c383215c03ce25a7df85ec7796e04f8
ca4723 :: /home/jovyan
```

You can now access your semantic_analysis container using the URL of http://loc
alhost:32776/?token=d57ccd98004c383215c03ce25a7df85ec7796e04f8ca4723.

Extending the Jupyter Image Using conda Environments

The original tool for managing Python environments is the virtualenv tool, still widely used today. Both virtualenv and the conda environments allow the user to create isolated Python environments on the same system, similar to the Bundler tool for Ruby programming. The Jupyter Docker stacks use conda environments to manage environments and you will leverage the existing environments to install packages for a Python 2 kernel independent of the installation of packages for a Python 3 kernel and vice versa.

You can manually switch back and forth between the two using the source activate command. This is done automatically at kernel launch by the notebook server, where necessary. In Listing 7-18, you examine the conda environments on the jupyter/scipy-notebook image.

■ **Note** base-notebook and minimal-notebook only have Python 3, but scipy-notebook and all child images have both Python 2 and Python 3.

Listing 7-18. Display Location and Version of python Binaries for Default, root, and python2 Environments

```
$ docker run -it --rm jupyter/scipy-notebook bash
jovyan@83a2aec2da92:~$ which python
/opt/conda/bin/python
jovyan@83a2aec2da92:~$ python --version
Python 3.5.2 :: Continuum Analytics, Inc.
jovyan@83a2aec2da92:~$ source activate root
(root) jovyan@83a2aec2da92:~$ python --version
Python 3.5.2 :: Continuum Analytics, Inc.
(root) jovyan@83a2aec2da92:~$ source activate python2
(python2) jovyan@83a2aec2da92:~$ python --version
Python 2.7.12
```

A proper extension of a base jupyter image would do so for each of the conda environments available on the image, as seen in Listing 7-19. Here, you are installing the pymongo library for both conda environments.

Listing 7-19. A Proper Extension of the jupyter/scipy-notebook Image

```
FROM jupyter/scipy-notebook

USER root

# python 3 environment is named root
RUN conda install --name root \
    pymongo

# python 2 environement is named python2
RUN conda install --name python2\
    pymongo

USER jovyan
```

■ **Note**　You switch to the user root to install the libraries and switch back to user jovyan upon completion. This is considered a best practice and ensures that you do not run the notebook server with too much system privilege.

Once more you build the image using the docker build command (Listing 7-20) and run a container defined by this image using docker run (Listing 7-21). It is not necessary to give the image a new name. You simply overwrite the previous semantic_analysis image.

Listing 7-20. Build the semantic_analysis Image

```
$ docker build -t semantic_analysis .
```

Listing 7-21. Run the semantic_analysis Container

```
$ docker run -d -P semantic_analysis
ca525dbeb79b8e38a52db192b7388136e61e4a1534817f76ba5eefd8ffc0246e
$ docker port ca52
8888/tcp -> 0.0.0.0:32777
$ docker exec ca52 jupyter notebook list
Currently running servers:
http://localhost:8888/?token=d57ccd98004c383215c03ce25a7df85ec7796e04f8
ca4723 :: /home/jovyan
```

You can now access your new semantic_analysis container using the URL http://
localhost:32777/?token=d57ccd98004c383215c03ce25a7df85ec7796e04f8ca4723.

One final consideration in extending jupyter images is that some python libraries
may only be available via pip. You may wish to install the python library twitter for
interfacing with the Twitter API. Again, you should install the library for both conda
environments, as you have done in Listing 7-22.

Listing 7-22. A Proper Extension of the jupyter/scipy-notebook Image using pip

```
FROM jupyter/scipy-notebook

USER root

# python 3 environment is named root
RUN conda install --yes --name root \
    pymongo

# python 2 environement is named python2
RUN conda install --yes--name python2\
    pymongo

# install libraries via pip using bash and activating respective environment
RUN ["bash", "-c", "source activate root && pip install twitter"]
RUN ["bash", "-c", "source activate python2 && pip install twitter "]

USER jovyan
```

Using joyvan to Install Libraries

In Listing 7-22, you make use of a variant to the Dockerfile RUN syntax. You have previously used the RUN instruction using the syntax RUN <command>. This is known as the *shell* form of the RUN instruction. Using the command in this form has the effect of sending the command to an image via /bin/sh or the default shell.

Installing via pip will not work using this syntax. This is because in order to install into the correct conda environment, you need to activate the appropriate environment using the source command. The source command is not available to the default shell (/bin/sh). It is available to bash. In order to use bash, you must use the alternative syntax for the RUN instruction, the *exec* form. The *exec* form uses the syntax RUN ["executable", "param1", "param2"]. You use RUN ["bash", "-c", "source activate <environment> && pip install twitter "]. This has the effect of opening a bash shell to your image and

1. Activating the appropriate environment

2. Installing twitter via pip

Ephemeral Container Extension

Finally, let's discuss a best practice in adding libraries to a running container using the jovyan user. Prior to doing so, however, I should mention the implications of making changes to a running container. In Chapter 2, you explored persistence in containers, but did so solely with regard to files you might be working on while using Jupyter. You wanted to make sure that changes to files persisted beyond the lifespan of a container. I did not, however, discuss persistence of libraries you might install, either at the system level using the apt package manager favored by the base-notebook's operating system Debian or at the Python level using any of the included python package managers.

In practice, you will face the exact same challenges. With regard to system and pythonic libraries, however, I recommend making these changes via a Dockerfile as a best practice, as described above. That said, it is also a best practice to separate the writing of python code from the maintenance of Docker images. You will, no doubt, in the course of development, come across a situation where you wish to install a library for quick use. A best practice is certainly to quickly install the needed library and continue developing.

You should be aware, however, of the implications of what you are doing. As soon as you stop and remove the container in which you have made these changes, the changes will be gone. This is by design. You must also guard yourself against keeping aging containers around long after their lifecycle because you wish to keep the changes you have made to them. With this in mind, I have found the following to be a best practice for installing **ephemeral** packages on running containers.

The ipython magix-sx command,[8] for which ! (bang) serves as a shorthand, can be used to run a command in a shell to the underlying system *from a notebook*. In this way, you can install libraries via conda (Listing 7-23), pip (Listing 7-25), or apt (Listing 7-27) in the moment, without opening a terminal to the container.

■ **Note**   apt requires that password-less sudo has been granted to jovyan at container runtime, as outlined above.

Listing 7-23. Install a Package via conda Using ipython Magic Shell Process Command.

```
In [1]: import pymongo
        --------------------------------------------------------------------
        ImportError                               Traceback (most recent
        call last)
        <ipython-input-1-ec8fdd1cd630> in <module>()
        ----> 1 import pymongo

        ImportError: No module named 'pymongo'
In [2]: !conda install pymongo --yes

        Fetching package metadata ........
        Solving package specifications: ..........

        ...

        Linking packages ...
        [      COMPLETE      ] |#####################################| 100%
```

The environment into which the library will be installed is a function of which kernel is being run. Figure 7-4 shows the Python 3 kernel. Recall that this is associated with the conda root environment.

---

[8]https://ipython.org/ipython-doc/3/interactive/magics.html#magic-sx

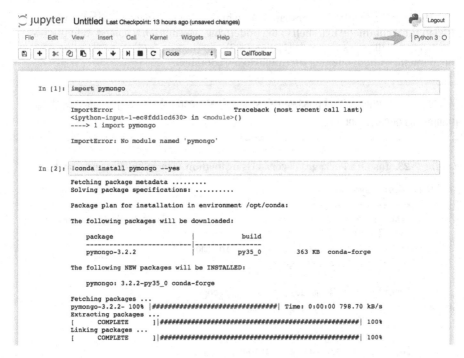

Figure 7-4. *Install pymongo into the root environment using a notebook*

You can verify that the package has indeed been added to the correct environment by opening a shell to the container and running a conda list (Listing 7-24).

Listing 7-24. Verify Library Installation

```
$ docker exec -it kickass_engelbart bash
jovyan@67cf8215ed6a:~$ conda list --name root | grep pymongo
pymongo                   3.2.2                    py35_0    conda-forge
jovyan@67cf8215ed6a:~$ conda list --name python2 | grep pymongo
jovyan@67cf8215ed6a:~$
```

The library is available to the root environment but it is not available to the python2 environment. Were you to switch kernels and run the command once more, it would be available to python2. These changes to the container will persist as long as the container is running. If you wish for these changes to persist for more than the most trivial usage, it is recommended to make these changes permanent by making them to the image itself, as outlined above.

In Listing 7-25, you install twitter into the python2 environment using pip via a notebook. See Figure 7-5 for the results.

Listing 7-25. Install a Package via `pip` Using ipython Magic Shell Process Command

```
In [1]: import twitter
        ---------------------------------------------------------------
        ImportError                             Traceback (most recent
                                                call last)
        <ipython-input-1-645f6dc1896f> in <module>()
        ----> 1 import twitter

        ImportError: No module named 'twitter'
In [2]: !pip install twitter

        Collecting twitter
          Downloading twitter-1.17.1-py2.py3-none-any.whl (55kB)
            100% |████████████████████████████████| 61kB
            2.4MB/s ta 0:00:01
        Installing collected packages: twitter
        Successfully installed twitter-1.17.1
```

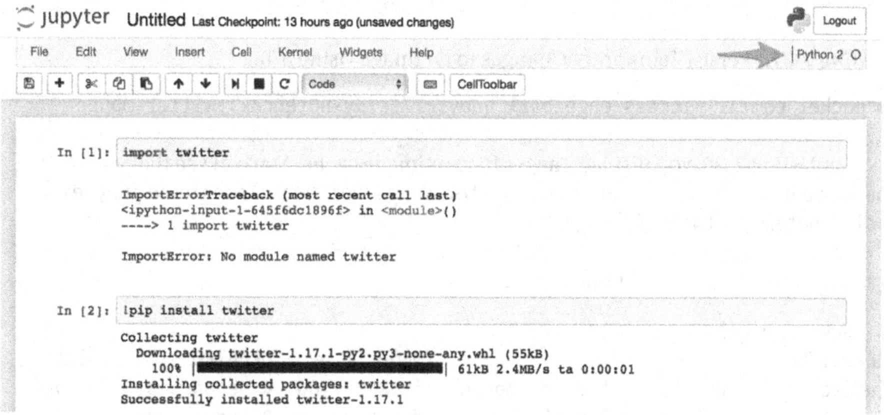

Figure 7-5. *Install* `twitter` *into the python2 environment using a notebook*

Again, you can verify that the package has indeed been added to the correct environment by opening a shell to the container and running a conda list (Listing 7-26).

Listing 7-26. Verify Library Installation.

```
$ docker exec -it kickass_engelbart bash
jovyan@67cf8215ed6a:~$ conda list --name root | grep twitter
jovyan@67cf8215ed6a:~$ conda list --name python2 | grep twitter
twitter                    1.17.1                    <pip>
jovyan@67cf8215ed6a:~$
```

133

Maintaining Semi-Persistent Changes to Images

The DevOps-minded reader will no doubt be troubled by the installation of libraries in a running Jupyter container. You have thus far clung to the notion that the best practice in working with containers is that they can be shut down, removed, and restarted at any time without any loss to your workflow. Were you to reap kickass_engelbart and start your notebook in a new container, this new container would not have twitter or mongo installed. This presents only a mild inconvenience as you have a record of libraries you will need to install in the magix-sx command in your notebooks: !conda install pymongo --yes and !pip install twitter. You may wish, however, to persist these changes in a defined container.

■ **Warning**   This practice is not recommended as a permanent image development practice. This is recommended as a temporary measure on a project-by-project basis.

You can add temporary changes (Listing 7-27) made to an image using the docker commit command. I recommend using a tag to name the image. Here you add the twitter-mongo tag to the base image, jupyter/scipy-notebook.

Listing 7-27. Persist Temporary Changes to an Image Using a Tag

```
$ docker commit kickass_engelbart jupyter/scipy-notebook:twitter-mongo
```

In Listing 7-28, you display images in your image cache. You can see that you now have two images in your local cache, jupyter/scipy-notebook:latest and jupyter/scipy-notebook:twitter-mongo.

Listing 7-28. Display Local Images

```
$ docker images
REPOSITORY              TAG             IMAGE ID      CREATED        SIZE
jupyter/scipy-notebook  twitter-mongo   dfbb7599770d  4 seconds ago  5.37GB
jupyter/scipy-notebook  latest          3dc12029099d  24 hours ago   5.35GB
```

You can launch your modified image as any other image using docker run (Listing 7-29).

Listing 7-29. Launch jupyter/scipy-notebook:twitter-mongo

```
$ docker run -d -v `pwd`:/home/jovyan -p 8888:8888 jupyter/scipy-notebook:twitter-mongo
```

Summary

In this chapter, you looked at the Docker images defined by the Jupyter team for quick launch and extension of the Jupyter Notebook server using Docker. I briefly discussed default notebook security strategies. I also presented several strategies for extending Jupyter images using the Jupyter Team's images as base. Following this chapter, I hope you feel comfortable accessing the right Jupyter image for your work and adding to this image to meet the needs of a given project.

CHAPTER 8

■ ■ ■

The Data Stores

In this chapter, let's extend the discussion beyond the Jupyter Notebook server to explore open source data store technologies and how we can use Docker to simplify the process of working with these technologies. I propose that using Docker, it is possible to streamline the process to an extent that using a data store for even the smallest of datasets becomes a practical matter. I'll show you a series of best practices for designing and deploying data stores, a set of practices that will be sufficient for working with all but the largest of data sets. Conforming to Docker best practice, you will work with Docker Hub official images throughout this chapter.

You will look at three data store technologies here: Redis, MongoDB, and PostgreSQL. Deploying and using these three technologies as services managed by Docker will require you to pay specific attention to two things:

1. Managing persistence in Docker containers using volumes

2. Networking between Docker containers

With regard to data persistence, you will establish a simple-but-effective best practice using Docker volumes that will be repeated for each data store. With regard to networking, you will explore three different options: legacy links, deploying a service on its own AWS instance, and manually creating bridge networks. You will look at legacy links for running Redis, AWS for running Mongo, and manual network creation for running PostgreSQL, but to be clear, any of these techniques could be applied to any of the data stores. I am merely presenting them in this fashion for learning purposes. It is recommended that you absorb all three networking techniques and choose the best option for each project you find yourself working on.

You will be sourcing each of the data stores from their respective Docker Hub community pages.

Serialization

A central task in the workflow of any data scientist is the storage, transmission, and reconstruction of data structures and object states. This process is known as serialization. It is a well-solved problem and you will have several tools at your disposal to manage this task. In this chapter, you will look at serialization in terms of converting objects in memory to their binary representation as well as the use of the popular JSON format for serialization as a text file. In Chapter 9, you will see a second format of text file serialization, the YAML format.

© Joshua Cook 2017
J. Cook, *Docker for Data Science*, DOI 10.1007/978-1-4842-3012-1_8

You will be serializing and deserializing primarily for the purposes of sharing objects and data across processes. In particular, you will be storing data in your databases, as well as caching objects in Redis for the purposes of using them in a separate notebook or process. Later, in Chapter 9, you will begin to define application architecture using code. In this case, you will be serializing your application configuration.

Serialization Formats and Methods

This book places an emphasis on working in Python. As such, it will focus on two Python-specific methods for serializing data: pickling and serializing via bytestring. In addition, you will look at appropriate uses for two text-based approaches to serializing data: JSON and YAML. JSON (JavaScript Object Notation) is a machine-readable subset of the JavaScript programming language that has been adopted by the programming community as a human readable, language agnostic approach to serialization. YAML is an alternative solution to the exact same problem. Both JSON and YAML are able to use the standard primitive data types: integers, floating-point numbers, Booleans, and null values, in addition to strings. For providing larger structures, both make use of the associative array, often called the dictionary, and the ordered list, also known as the array, the vector, the list, or the sequence. A dictionary holds data using key-value pairs; a list holds data using a numerical index. The two mainly differ in syntax. JSON (Listing 8-1) makes use of nested braces and brackets to define data structures, while YAML (Listing 8-2) achieves the same purpose using white space. Note that in the following two examples, none of the keys used have any syntactical meaning.

Listing 8-1. A Sample JSON Object

```
{'this_json' : 'is a JSON object',
 'a nested object' : {
  'obj_id' : 123,
  'object value' : 'temperamental',
  'is_nested' : true
  },
 'a list': [1,2,3,4],
 'a list of strings': ['green eggs', 'ham'],
 'last_used' : null
}
```

Listing 8-2. A Sample YAML Object

```
this_yaml: is a YAML object
a_nested_object:
  obj_id: 123
  object_value: 'temperamental'
  is_nested: true
a_list:
  - 1
  - 2
```

```
    - 3
    - 4
a_list_of_strings:
    - green eggs
    - ham
last_used: null
```

Binary Encoding in Python

The Python pickle module is the preferred method for serialization of Python objects and data to binary byte streams. There are a few fundamental differences between pickling data and serializing using JSON or YAML. As noted, both JSON and YAML are human readable. An object converted to a byte stream is not human readable. JSON and YAML serialized objects will be readable by a process run in any language, while a pickled object will only be readable in Python. Because a pickled object does not have to be concerned with interoperability, a wide variety of Python objects can be pickled, whereas only dictionaries can be serialized using JSON or YAML. For the data scientist, this includes but is not limited to the numpy array, the pandas DataFrame, or the sklearn Model. Over the next chapter, you will explore a variety of methods for encoding data to a binary byte stream using Python.

Redis

Redis[1] is an open source, in-memory data structure store. It stores data values of several different types associated to a given key. In your stack, you will use Redis for two purposes. First, it will serve as a cache for persisting objects beyond the lifespan of a Python process or in-between Jupyter Notebooks. Second, you will use Redis as a message broker in order to perform delayed job processing from your notebooks using the Python library named rq. In this chapter, you will address the first use case; a discussion of the second use case happens in Chapter 10.

Pull the redis Image

You can retrieve the redis image from the Docker Hub using docker pull, as in Listing 8-3. The Docker Hub page[2] for the redis image outlines much of what I will discuss in the next few pages including basic configuration, persistent storage, and connecting to Redis from another container.

---

[1]https://redis.io
[2]http://hub.docker.com/_/redis

Listing 8-3. Pull the redis Image from Docker Hub

```
$ docker pull redis
Using default tag: latest
latest: Pulling from library/redis
6d827a3ef358: Already exists
787f13ab8ea9: Pull complete
...
Digest: sha256:1b358a2b0dc2629af3ed75737e2f07e5b3408eabf76a8fa99606ec0c276a93f8
Status: Downloaded newer image for redis:latest
```

■ **Note** You already had the first layer of the redis image in your local Docker image cache and the pull command notifies you that layer 6d827a3ef358 "Already exists". This is because Redis uses the same base image as Jupyter, that is, they both begin their respective Dockerfiles with FROM debian:jesse.

You can minimally verify that the image functions by running a Redis container in detached mode (-d) (Listing 8-4) and sending a ping command to the server via the redis-cli (Listing 8-5). To interface with Redis via the redis-cli you need to issue the command via a docker exec statement issued to the proper container. When the Redis server running in the container responds with PONG, you know that all is good.

Listing 8-4. Run a Redis Container in Detached Mode

```
$ docker run -d redis
96d6ddb6d06f1422b11193ac84a18346f3be53fd7912dc38b6301c0573171647
```

In Listing 8-5, you issue a command to the running Redis container via docker exec, making reference to the running container by the first four characters of its container id, 96d6.

Listing 8-5. Ping the Redis Server

```
$ docker exec 96d6 redis-cli ping
PONG
```

Finally, you shut down and remove your running Redis container (Listing 8-6).

Listing 8-6. Shut Down and Remove the Running Redis Container

```
$ docker stop 96d6
96d6
$ docker rm 96d6
96d6
```

Docker Data Volumes and Persistence

In thinking about running the data stores using Docker, it is helpful to think of them as services (or microservices) that are being managed by Docker. Recall that containers themselves should be ephemeral. You should be able to start, stop, and discard containers at will. If the data in your data store is saved within the container, it will be lost each time a container is cycled. Just as when running a Jupyter Notebook server, persistence of data beyond the lifespan of a container will be mission critical for all of your data store containers. Let's loosely define this idea of a "service" as a container **and** its persistent data.

■ **Note** It is worth emphasizing that I am loosely defining the idea of a "service." We are not working with the Docker tool `service`.

With notebooks, you persisted data by mounting a host directory as a data volume using the volume flag (`-v`) (Listing 3-32). With your data stores, you will address the issue by using data volumes.[3] A data volume is a specially designed container specifically designed to persist data beyond the lifespan of associated containers. Now you can think of a service as the container and its linked volume (Figure 8-1). Because the data volume persists beyond the lifespan of a single container, you can start, stop, and remove the container, and attach a new container to the existing data volume, without any loss of data.

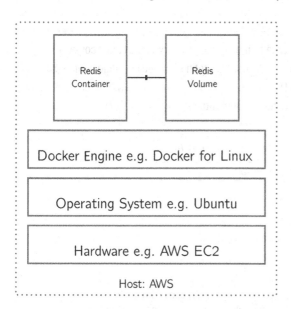

Figure 8-1. *Redis as a persistent service being managed by Docker*

---

[3]https://docs.docker.com/edge/engine/reference/commandline/volume/

Create and View a New Data Volume

First, you create a new data volume (Listing 8-7).

Listing 8-7. Create a New Redis Data Volume Container

```
$ docker volume create --name redis-dbstore
redis-dbstore
```

Volumes exist apart from containers and thus are viewed independently using the docker volume ls command (Listing 8-8).

Listing 8-8. View Current Docker Volumes

```
$ docker volume ls
DRIVER              VOLUME NAME
local               redis-dbstore
```

Launch Redis as a Persistent Service

You will connect to the newly created volume at runtime in order to allow your Redis container to persist its data beyond its lifespan. In addition to connecting to the data volume, you will run the Redis container in detached mode (-d) and give the container a name (--name) (Listing 8-9).

■ **Note**   You mount the docker volume to the /data directory in the Redis container. This location is specified in the Dockerfile used to define the redis image, which can be viewed on the Redis public page on Docker Hub.[4] Care must be taken with each data store, as each will be looking for its data cache in a separate location (Table 8-1).

Table 8-1. *Location of Data Cache Within Container by Image*

Image	Data Cache Loction
Redis	/data
Mongo	/data/db
Postgres	/var/lib/postgresql/data

Listing 8-9. Launch Redis with an Attached Volume

```
$ docker run -d --name this_redis -v redis-dbstore:/data redis
b216a67caedc934b09341cf1642e89079be09d52b607ce4ddecdeaae5b5ae704
```

---

[4]https://hub.docker.com/_/redis/

You can verify the persistence of data in the following manner. In Listing 8-10, you create a new incr object on your Redis server using the redis-cli, and ping the incr twice more for good measure.

Listing 8-10. Create a New incr Object via redis-cli

```
$ docker exec this_redis redis-cli incr mycounter
1
$ docker exec this_redis redis-cli incr mycounter
2
$ docker exec this_redis redis-cli incr mycounter
3
```

In Listing 8-11, you shut down and remove the container named this_redis.

Listing 8-11. Shut Down and Remove this_redis

```
$ docker stop this_redis && docker rm this_redis
this_redis
this_redis
```

Next, in Listing 8-12, you start up a new container using the same command as before. If the data has persisted, then a subsequent incr command will yield a 4 (Listing 8-13).

Listing 8-12. Start Up a New Redis Instance for Use

```
$ docker run -d --name this_redis -v redis-dbstore:/data redis
12fe7cea2e63aa2055585fd97b6b9205774a59bbf716672f71e5d75858c7cd72
```

Listing 8-13. Issue an incr Command via redis-cli

```
$ docker exec this_redis redis-cli incr mycounter
4
```

You received a response of 4 because the volume you have created and attached to each container in turn has allowed you to persist the Redis data between container instances. You can think of this as starting and stopping Redis as a native process that reconnects to a stored file on disk each time it runs.

Connecting Containers via Legacy Links

Having solved the persistence issue, let's now overcome the next hurdle: connecting to your container. The easiest way to connect to a container running on the same host machine is via the --link flag at runtime. This is done simply by adding the --link flag with a reference to a named running container to the docker run command issued to launch a new container. In Listing 8-14, you launch a new Jupyter container with a link to your named Redis container, this_redis, as visualized in Figure 8-2.

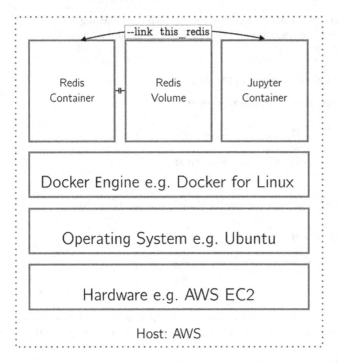

Figure 8-2. *Connecting Redis and Jupyter on the same system*

Listing 8-14. Launch a New Jupyter Container Linked to the Running Redis Container

```
$ docker run -d -v `pwd`:/home/jovyan --link this_redis jupyter/scipy-notebook
d6f09196bf85861df23eeb2f11bd68396287464d00febe27cda93024a3666251
```

It is worth the effort to spend a moment studying the form of the connection created by this link. You will do so by opening a bash shell to the running Jupyter container and examining the environment (Listing 8-15). Again, you use the shorthand of the first four characters of the container id d6f0 to facilitate your connection.

Listing 8-15. Explore the Environment of a Running Jupyter Container

```
$ docker exec -it d6f0 bash
jovyan@d6f09196bf85:~$ env | grep THIS_REDIS
THIS_REDIS_PORT_6379_TCP=tcp://172.17.0.2:6379
THIS_REDIS_NAME=/determined_wilson/this_redis
THIS_REDIS_PORT=tcp://172.17.0.2:6379
THIS_REDIS_PORT_6379_TCP_PORT=6379
THIS_REDIS_ENV_REDIS_VERSION=3.2.8
THIS_REDIS_PORT_6379_TCP_PROTO=tcp
THIS_REDIS_ENV_GOSU_VERSION=1.7
THIS_REDIS_ENV_REDIS_DOWNLOAD_SHA1=6780d1abb66f33a97aad0edbe020403d0a15b67f
THIS_REDIS_ENV_REDIS_DOWNLOAD_URL=http://download.redis.io/releases/redis-3.2.8.tar.gz
THIS_REDIS_PORT_6379_TCP_ADDR=172.17.0.2
```

Here, you use the env command line tool to display the defined environment. You then pipe (|) the output of the command to the grep tool matching on the pattern THIS_REDIS. Note that docker has defined numerous environment variables with the name THIS_REDIS. This environment variable will always be an all-caps version of the container's name. Your container is named this_redis, and thus the environment variables will use THIS_REDIS.

You will be able to use any of these environment variables to facilitate your connection to Redis rather than needing to take note of an IP address each time you run the container.

In Listing 8-16, you demonstrate this by sending a ping from the Jupyter container to the Redis container using the environment variable, THIS_REDIS_PORT_6379_TCP_ADDR. Note that you use the $ syntax to reference the environment variable.[5]

Listing 8-16. Ping Redis from Jupyter

```
jovyan@d6f09196bf85:~$ ping -c 4 $THIS_REDIS_PORT_6379_TCP_ADDR
PING 172.17.0.2 (172.17.0.2): 56 data bytes
64 bytes from 172.17.0.2: icmp_seq=0 ttl=64 time=0.470 ms
64 bytes from 172.17.0.2: icmp_seq=1 ttl=64 time=0.136 ms
64 bytes from 172.17.0.2: icmp_seq=2 ttl=64 time=0.120 ms
64 bytes from 172.17.0.2: icmp_seq=3 ttl=64 time=0.085 ms
--- 172.17.0.2 ping statistics ---
4 packets transmitted, 4 packets received, 0% packet loss
round-trip min/avg/max/stddev = 0.085/0.203/0.470/0.155 ms
```

■ **Warning** At the time of this writing, the --link flag is a deprecated legacy feature. Per the Docker documentation, "it may be eventually removed." For our purposes, it provides such a straightforward method for connecting containers that I think it worth discussing, even though it has been deprecated.

Using Redis with Jupyter

This minimal connection is sufficient to verify that your containers are able to communicate with each other, but you will need to be able to interface with the redis server from within a Jupyter Notebook. By default, the jupyter/scipy-notebook image does not include the Redis Python library necessary to interface with a Redis server from within a Python process. While it is best practice to define a new image that includes the libraries that you wish to use, here you will use an ephemeral installation of the Redis library inside a running container to demonstrate how you might quickly interface with the Redis server from within a notebook.

---

[5]http://tldp.org/LDP/abs/html/ivr.html

In Listing 8-17 and Figure 8-3, you can see that the container does not in fact have the Python library Redis, after which you install the library using pip executed in a subprocess from the Jupyter notebook.[6]

Listing 8-17. Install Redis via a Shell Call to a Jupyter Notebook

```
In [1]: import redis
        ---------------------------------------------------------------
        ImportError                        Traceback (most recent call last)
        <ipython-input-1-6872e27f77ac> in <module>()
        ----> 1 import redis

        ImportError: No module named 'redis'
In [2]: !pip install redis
        Collecting redis
          Downloading redis-2.10.5-py2.py3-none-any.whl (60kB)
            100% |████████████████████████████████| 61kB 2.5MB/s ta 0:00:01
        Installing collected packages: redis
        Successfully installed redis-2.10.5
        You are using pip version 8.1.2, however version 9.0.1 is available.
        You should consider upgrading via the 'pip install --upgrade pip' command.
```

Figure 8-3. Install Redis via a shell call from a Jupyter Notebook

---

[6]http://ipython.readthedocs.io/en/stable/interactive/python-ipython-diff.html?highlight=!ls#quick-overview

A Simple Redis Example

Having configured the `jupyter/scipy-notebook` container with the Python Redis library, you are ready to begin using the service. You can think of Redis as giving you the ability to read and write values to RAM, allowing you to persist these values beyond the lifespan of a process or between currently running processes. Each of the examples below demonstrates this by using Redis to share values between two running Jupyter Notebooks.

Figure 8-4 demonstrates connecting two separate notebooks to the Redis service using the pattern outlined in Listing 8-18.

Listing 8-18. Connect to the Redis Service

```
In [1]: from redis import Redis
        from os import environ
        REDIS = Redis(host=environ['THIS_REDIS_PORT_6379_TCP_ADDR'])
```

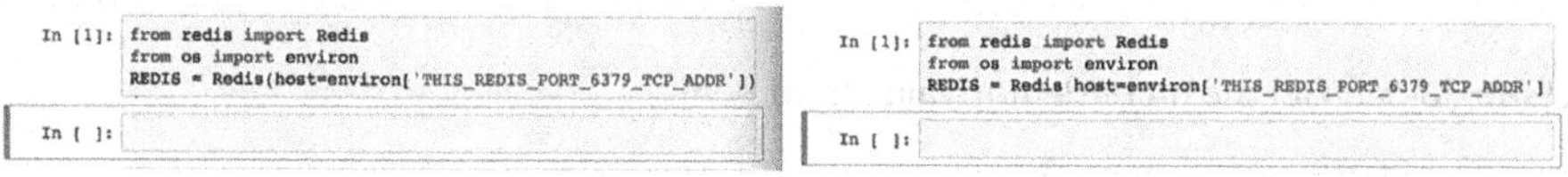

Figure 8-4. *Instatiate a Redis connection from two separate notebooks*

Next, in Figure 8-5, you set a key-value pair in Redis from the left notebook and retrieve the value in the right notebook, using the code patterns in Listing 8-19 and 8-20, respectively.

Listing 8-19. Set a Key-Value Pair in Redis

```
In [2]: REDIS.set('foo', 42)
```

Listing 8-20. Get a Value from Redis

```
In [2]: REDIS.get('foo')
Out[2]: b'42'
```

Figure 8-5. *Store a key-value pair from one notebook to be retrieved by another*

The power of what you have done here may not at first be obvious. You have shared a value from one notebook to another (i.e. one python process to another). Such a tool can be extraordinarily valuable.

Track an Iterative Process Across Notebooks

In Figure 8-6, you see how Redis can be used to track the development of an iterative process. You use the code pattern outlined in Listing 8-21 to mock the execution of an iterative process and the code pattern in Listing 8-22 to check the progress of the process while it is in the midst of execution.

Listing 8-21. Mock an Iterative Process

```
In [3]: import time
        def some_iterative_process():
            time.sleep(1)
In [4]: count = 0
        REDIS.set('count', 0)

        while count < 30:
            some_iterative_process()
            count = REDIS.incr('count')
```

Listing 8-22. Get the Incrementor Value from Redis

```
In [2]: REDIS.get('count')
Out[2]: 8
```

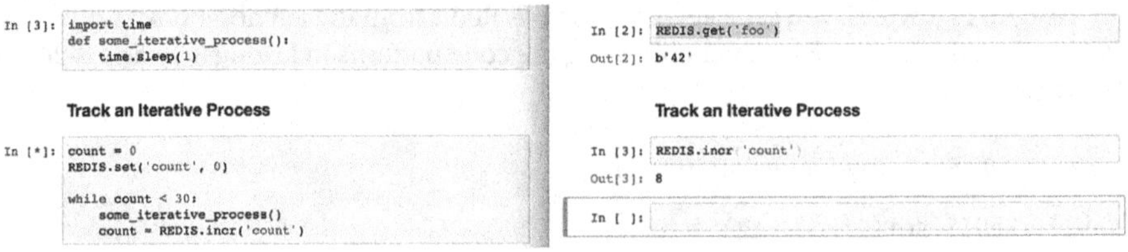

Figure 8-6. *Track an iterative process across notebooks*

The In [*] on the left notebook indicates that the process is currently running. You obtain the current count in the right notbook by retrieving the count from Redis. The count is continually updated at each iteration by the process being run on the right.

Pass a Dictionary via a JSON Dump

You might also pass a dictionary of values from one notebook to another using Redis. Figure 8-7 demonstrates this with a dictionary containing potential model parameters. You use the code pattern in Listing 8-23 to define the dictionary and pass it to Redis. You use the pattern in Listing 8-24 to first load the object from Redis and then to convert it to a Python dictionary.

Listing 8-23. Define a Dictionary and Pass It to Redis

```
In [5]: import numpy as np
        import json
        model_params = {
            'C': list(np.logspace(-3,3,7)),
            'penalty': 'l1',
            'solver' : 'newton-cg'
        }

        REDIS.set('model_params', json.dumps(model_params))
Out[5]: True
```

Listing 8-24. Load a Dictionary from Redis

```
In [4]: REDIS.get('model_params')
Out[4]: b'{"C": [0.001, 0.01, 0.1, 1.0, 10.0, 100.0, 1000.0],
        "solver": "newton-cg", "penalty": "l1"}'
In [5]: import json
        json.loads(REDIS.get('model_params').decode())
Out[5]: {'C': [0.001, 0.01, 0.1, 1.0, 10.0, 100.0, 1000.0],
        'penalty': 'l1',
        'solver': 'newton-cg'}
```

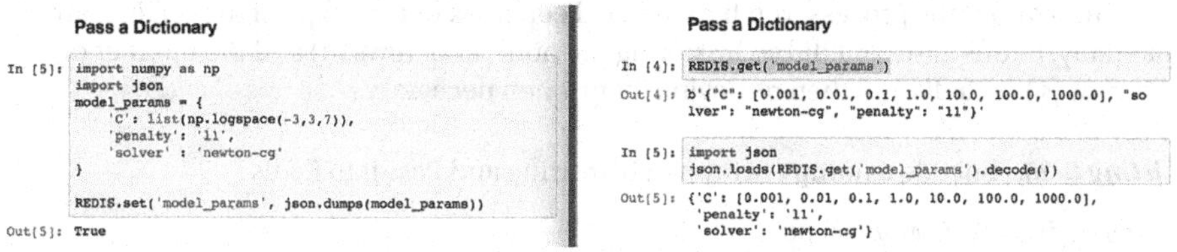

Figure 8-7. *Pass a dictionary via a JSON dump*

In the receiving notebook, when you initially load the object from Redis, it is returned as a bytestring, as is denoted by the leading b when the string is displayed in Out[4]. Most objects returned by Redis will be returned as bytestrings, and you must take special care to handle them. Here, you wish the object to be a dictionary. To have the object load as a dictionary, you first use the .decode() function included as part of the Python bytes class. This converts the object to a string. You then pass this string to the json.loads() function, which converts the string to the dictionary you see displayed as Out[5].

Pass a Numpy Array as a Bytestring

numpy's natural capacity for working with bytestrings makes it an excellent partner for Redis and it is very straightforward to pass numpy arrays and vectors. In Listings 8-25 and 8-26, you encode and store, and then load and decode, a numpy array.

Note that in order to convert an array to a bytestring, you must first convert it to a vector. This is done using the .ravel() function included as part of the np.array class. This function converts an array of shape (n, m) to a vector of shape (n*m,). You then convert the vector to a bytestring using the .tostring() function.

To convert from a bytestring, you use the numpy function np.fromstring() before using the .reshape() function to convert the resulting vector back into an array. See Figure 8-8.

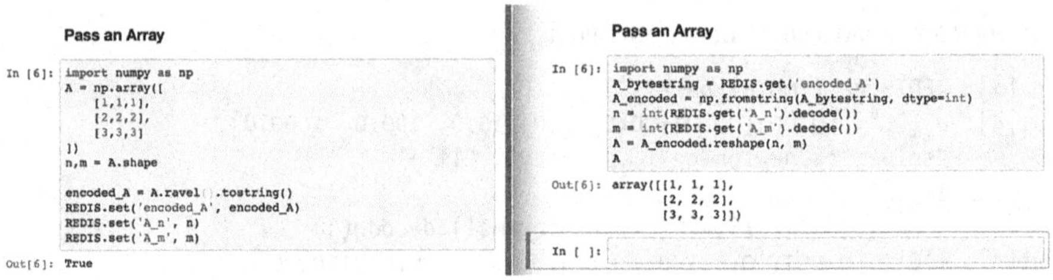

Figure 8-8. *Pass a numpy array as a bytestring*

Throughout the process, you have had to keep track of the shape of the original array manually. In this case, you did so by storing the number of rows (n) and the number of columns (m) in Redis and then retrieving them when necessary.

Listing 8-25. Encode a numpy Array as a bytestring and Pass it to Redis

```
In [6]: import numpy as np
        A = np.array([
            [1,1,1],
            [2,2,2],
            [3,3,3]
        ])
        n,m = A.shape

        encoded_A = A.ravel().tostring()
        REDIS.set('encoded_A', encoded_A)
        REDIS.set('A_n', n)
        REDIS.set('A_m', m)
Out[6]: True
```

Listing 8-26. Load a numpy Array and Decode

```
In [6]: import numpy as np
        A_bytestring = REDIS.get('encoded_A')
        A_encoded = np.fromstring(A_bytestring, dtype=int)
        n = int(REDIS.get('A_n').decode())
        m = int(REDIS.get('A_m').decode())
        A = A_encoded.reshape(n, m)
        A
Out[6]: array([[1, 1, 1],
               [2, 2, 2],
               [3, 3, 3]])
```

■ **Note**   By default, np.fromstring() will convert a passed bytestring to a vector of floating point values. Since you passed a vector of integers, you must specify this type when converting from bytestring, as you have done with np.fromstring(A_bytestring, dtype=int).

MongoDB

Redis is a data structure store. It is largely concerned with storing individual object instances. MongoDB is a database. It is concerned with storing many instances of the same type. MongoDB calls each instance that it stores a *document* and stores these documents using a JSON-like format. This format also lends itself to working with the Python dictionary class.

As compared to other databases, MongoDB is known for its flexibility. Rather than dealing with complex schema in order to store object instances, you can quickly and easily add an object to a MongoDB collection without concerning yourself with the very idea of schema. Such an approach to databases is known as NoSQL, and MongoDB is perhaps the most widely used NoSQL database.

With regard to data persistence, you will use the same approach you used in setting up Redis. With regard to networking, you will take a slightly different approach. Rather than configuring MongoDB to run alongside a Jupyter service, you will configure Mongo to run on an AWS instance. The advantages of this come down to networking.

If you expose the necessary port in the Mongo container to the corresponding port on the host container, you no longer need to worry about networking. You simply connect to the IP address of the AWS machine. I have found this solution to be in practice slightly easier than configuring a bridge network to connect two containers on the same host system and does not use the now deprecated --link method previously discussed.

■ **Note**   As the recommended practice was for you to follow along in this book using a t2.micro, you may currently be using an AWS instance to do the work in this book. This is by design and I hope that this is still the case. The intent of the following section is for you to run MongoDB on a system that is apart from the one on which you are currently working. If you are currently working on an AWS instance, this means that you will now be working on two AWS instances. You will have these two instances communicate with each other over the open internet.

Set Up a New AWS t2.micro

You did this work earlier in Chapter 1, so I will merely outline the steps necessary to bring a new t2.micro online:

1. From the AWS EC2 Dashboard, select "Launch Instance."

2. On the Choose AMI tab, choose Ubuntu Server 16.04.

3. On the Choose Instance Type tab, choose t2.micro.

4. On the Add Storage tab, use the default setting of 8GB.

5. On the Configure Security Group tab, choose "Create a **new** security group."

    a. Confirm that inbound SSH traffic can be accepted over port 22 from anywhere.

    b. Add a rule that accepts inbound traffic over port 2376 from anywhere. This port will allow you to pull images from Docker Hub.

    c. Add a rule that accepts inbound traffic over port 27017 from anywhere. This is the default port for accessing MongoDB.

6. Review and launch an instance, taking care to confirm that you have access to the SSH keys stored with your AWS account.

■ **Note**   You have added port 27017 to this Security Group so that you can access MongoDB on its default port over the open web. If Redis is configured on its own AWS instance, then port 6379 will need to be added to the Security Group. If PostgreSQL is configured on its own AWS instance, then 5432 will need to be added to the Security Group.

Configure the New AWS t2.micro for Docker

As before, you provision the new instance with Docker in order to do your work. Again, I will outline steps without a great deal of explanation. Code for this configuration is provided in Listing 8-27. Readers seeking additional information are referred to the detailed description of this process in Chapter 1.

1. Take note of the IP address of the newly configured AWS instance.

2. SSH into the instance using that IP address.

3. Install Docker via a shell script.

4. Add the ubuntu user to the docker group.

5. Log out and back in.

Listing 8-27. Provision the AWS Instance with Docker

```
(local) $ ssh ubuntu@255.255.255.255
(remote) $ curl -sSL https://get.docker.com/ | sh
(remote) $ sudo usermod -aG docker  ubuntu
```

After this last command, log out of the ssh session to AWS and log back in to configure MongoDB.

Pull the mongo Image

As with the redis image, you will retrieve the mongo image from its public page on Docker Hub.[7] The page also provides details about configuring Mongo. For your purposes, you are most interested in reading about the specifics of the default configuration. In Listing 8-28, you pull the image.

Listing 8-28. Pull the mongo Image

```
(local) $ ssh ubuntu@255.255.255.255
(remote) $ docker pull mongo
```

Create and View a New Data Volume

As with Redis, you will persist the data associated with your MongoDB by creating a data volume that will live beyond a container running Mongo. Again, you can think of MongoDB as a "service" comprised of the container running Mongo and the associated data volume (Figure 8-9). To set this up, you simply create a data volume for your Mongo data, as outlined in Listing 8-29.

---

[7]https://hub.docker.com/_/mongo/

Listing 8-29. Create and View a New Mongo Data Volume

```
$ docker volume create --name mongo-dbstore
$ docker volume ls
DRIVER              VOLUME NAME
local               mongo-dbstore
local               redis-dbstore
```

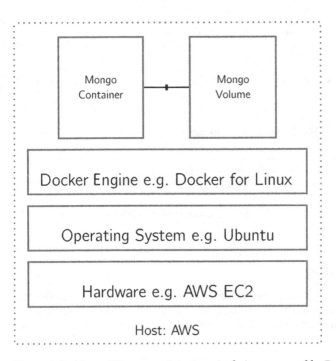

Figure 8-9. MongoDB as a persistent service being managed by Docker

Launch MongoDB as a Persistent Service

You connect to the newly created volume at runtime in order to allow your Mongo container to persist its data beyond its lifespan (Listing 8-30).

■ **Note**   You mount the docker volume to the /data/db directory in the Mongo container. This location is specified in the Dockerfile used to define the mongo image, which can be viewed on the Mongo public page on Docker Hub.[8]

---

[8]https://hub.docker.com/_/mongo/

Listing 8-30. Launch mongo with an Attached Volume

```
$ docker run -d --name this_mongo -v mongo-dbstore:/data/db -p 27017:27017 mongo
38a2f19d72a09851dc32cb874817a45274e888dd93aca01b5500cbfe9fb9364c
```

Verify MongoDB Installation

You can verify that you are running the mongo service by connecting to the running MongoDB via the MongoDB client, mongo, issued via docker exec (Listing 8-31).

■ **Warning** Both the image and the client share the same name, mongo. It is important to keep track of which is being referred to as you do your work. In the issuing of the command in Listing 8-31, mongo refers to the command line client you are using to interface with the running MongoDB.

In Listing 8-31, you connect and then insert a trivial document to a mongo collection. You are inserting the JSON object {"foo":1} into the collection test. You then search for the document you inserted using the .find() command.

Listing 8-31. Connect to mongo and Insert a Document

```
$ docker exec -it this_mongo mongo
MongoDB shell version v3.4.4
...
> db.test.insert({"foo":1})
WriteResult({ "nInserted" : 1 })
> db.test.find()
{ "_id" : ObjectId("591a00ee33e4717a80d8c92d"), "foo" : 1 }
```

Using MongoDB with Jupyter

As with Redis, you will need to be able to connect to MongoDB from within a Jupyter notebook. As before, you will need to install the necessary Python library, pymongo, and you will do this via a pip install issued through a notebook (Figure 8-10, Listing 8-32).

Listing 8-32. Install pymongo via a shell Call from a Jupyter Notebook

```
In [1]: !pip install pymongo
        Collecting pymongo
          Downloading pymongo-3.4.0-cp35-cp35m-manylinux1_x86_64.whl (359kB)
            100% |████████████████████████████| 368kB 700kB/s ta 0:00:01
        Installing collected packages: pymongo
        Successfully installed pymongo-3.4.0
```

```
In [1]: !pip install pymongo
        Collecting pymongo
          Downloading pymongo-3.4.0-cp35-cp35m-manylinux1_x86_64.whl (359kB)
            100% |████████████████████████████████| 368kB 700kB/s ta 0:00:01
        Installing collected packages: pymongo
        Successfully installed pymongo-3.4.0
        You are using pip version 8.1.2, however version 9.0.1 is available.
        You should consider upgrading via the 'pip install --upgrade pip' command.
```

Figure 8-10. *Install pymongo via a shell call from a Jupyter Notebook*

MongoDB Structure

As a NoSQL database, MongoDB has a minimal approach to structure. MongoDB has three kinds of entities:

1. The database

2. The collection

3. The document

These entities relate to each other in that databases hold collections which hold documents (Figure 8-11). Each document[9] is a binary representation of a JSON data record. Documents are composed of key-value pairs where a value can be any of the BSON data types.[10]

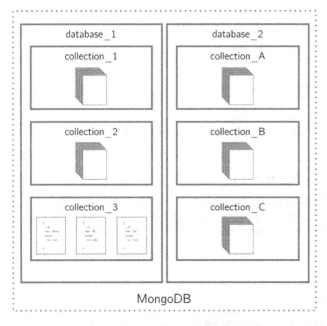

Figure 8-11. *MongoDB hierarchy*

---

[9]https://docs.mongodb.com/manual/core/document/#bson-document-format
[10]https://docs.mongodb.com/manual/reference/bson-types/

pymongo

pymongo[11] is a Python module containing the MongoDB tools recommended for working with the database. You begin (Listing 8-33) by instantiating a connection to MongoDB using pymongo.MongoClient. Here, you use the IP address of your AWS instance on which MongoDB is running.

Listing 8-33. Connect to MongoDB

```
In [2]: from pymongo import MongoClient
        client = MongoClient('255.255.255.255', 27017)
```

■ **Note** If using the default port of 27017 as you have done (Listing 8-30), it is not necessary to specify the port.

---

pymongo has a very useful "get or create" mechanism for both databases and collections. Databases and collections are accessed using either attribute-style (client.database_name) or dictionary-style (client['test-database']). If the database exists, this method will return a reference to the existing database or collection ("get"). If the database does not exists, this method will create the database or collection and then return a reference to it ("create"). The creation happens at the time of insertion of a document.

In Listing 8-34, you display currently extant databases.

Listing 8-34. Display Databases

```
In [3]: client.database_names()
Out[3]: ['admin', 'local', 'test']
```

Next, in Listing 8-35, you create a reference to a new database and once more show databases. Note that the database named my_database does not yet exist.

Listing 8-35. Create Database Reference and Display Databases

```
In [4]: db_ref = client.my_database
        client.database_names()
Out[4]: ['admin', 'local', 'test']
```

Next, in Listing 8-36, you create a reference to a new collection in my_database and then show databases, as well as collections, associated with my_database. Note that the database my_database still does not yet exist.

---

[11]http://api.mongodb.com/python/current/

Listing 8-36. Create Collection Reference and Display Databases and Collections

```
In [5]: coll_ref = db_ref.my_collection
        client.database_names(), db_ref.collection_names()
Out[5]: (['admin', 'local', 'test'], [])
```

In Listing 8-37, you create a new Python dictionary and insert the dictionary into your collection using the .insert_one() class function.

Listing 8-37. Insert a Document into a Collection

```
In [6]: sample_doc = {"name":"Joshua", "message":"Hi!", 'my_array' :
[1,2,3,4,5,6,7,9]}
        coll_ref.insert_one(sample_doc)
Out[6]: <pymongo.results.InsertOneResult at 0x7efc749726c0>
```

In Listing 8-38, you show the databases as well as the collections associated with my_database. As can be seen, now your database and collection both exist as they were created on insertion.

Listing 8-38. Display Databases and Collections

```
In [7]: client.database_names(), db_ref.collection_names()
Out[7]: (['admin',
         'local',
         'my_database',
         'test'],
        ['my_collection'])
```

In Listing 8-39, you demonstrate an interesting behavior of MongoDB. You drop all elements from my_collection using the .drop() class function and then show the databases and the collections associated with my_database once more. Note that neither my_database nor my_collection continue to exist. In other words, databases and collections exist solely as containers for documents.

Listing 8-39. Drop All my_collection Documents and Display Databases and Collections

```
In [8]: my_collection.drop()
        client.database_names(), db_ref.collection_names()
Out[8]: (['admin', 'local', 'test'], [])
```

Mongo and Twitter

To demonstrate a simple usage for MongoDB with Jupyter, you will implement a basic Twitter streamer that inserts captured tweets into a MongoDB collection. Twitter data represents an ideal use case for the NoSQL MongoDB. Each tweet obtained via the Twitter API is received as an unstructured nested JSON object. Adding such an object to a SQL database would be a non-trivial task by any measure involving numerous foreign keys and JoinTables as the user seeks to manage each of the one-to-one, one-to-many, and many-to-one relationships built into the tweet. Adding such an object to Mongo, on the other hand, is a trivial task. MongoDB's native Binary JSON (BSON) format was designed precisely to accept such an object.

Obtain Twitter Credentials

In order to follow along, you must obtain API credentials for accessing the Twitter API. This is done by creating a Twitter application.

In order to do this, follow these steps:

1. Visit `https://apps.twitter.com` and sign in.

2. Choose "Create New App" (Figure 8-12).

3. Give the new app a name, description, and website. For your purposes, the values of these responses are irrelevant, although the website will need to have a valid URL structure.

4. Agree to the Developer Agreement and click "Create your Twitter Application" (Figure 8-13).

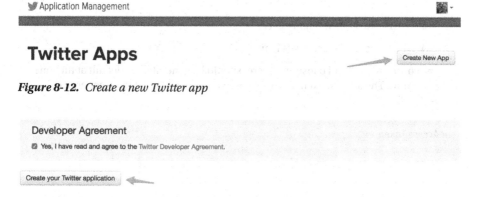

Figure 8-12. *Create a new Twitter app*

Figure 8-13. *Agree to the developer terms and create the app*

Once the new app is created, you will see details of your new app as shown in Figure 8-14.

Figure 8-14. *Your new Twitter app*

Next, you will need to access your credentials on the "Keys and Access Tokens" tab. You will need a total of four values:

1. A consumer key (API Key) (Figure 8-15)

2. A consumer secret (API Secret) (Figure 8-15)

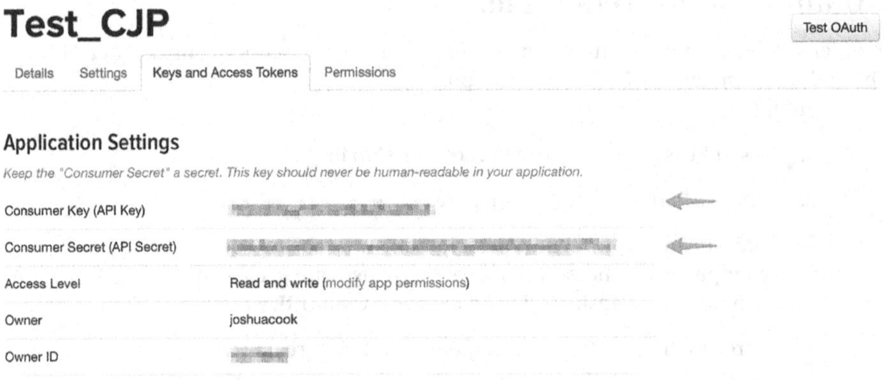

Test_CJP

Test OAuth

Details   Settings   **Keys and Access Tokens**   Permissions

Application Settings

Keep the "Consumer Secret" a secret. This key should never be human-readable in your application.

Consumer Key (API Key)	
Consumer Secret (API Secret)	
Access Level	Read and write (modify app permissions)
Owner	joshuacook
Owner ID	

Figure 8-15. Consumer key and consumer secret

3. An access token (Figure 8-17)

4. An access token secret (Figure 8-17)

The consumer key and consumer secret should be generated by default at the time of app creation. The access token and access token secret will need to be generated (Figure 8-16).

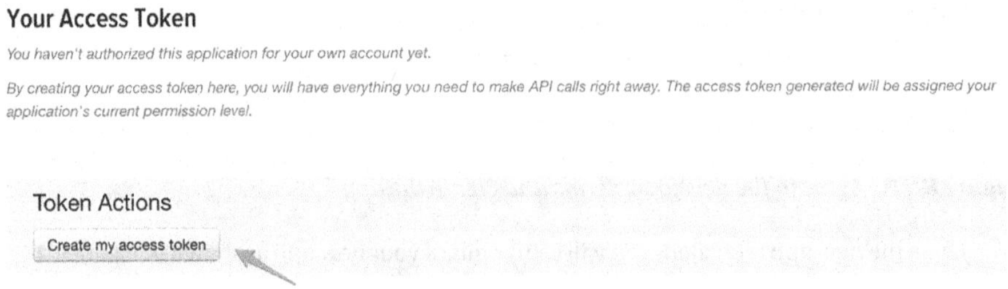

Your Access Token

You haven't authorized this application for your own account yet.

By creating your access token here, you will have everything you need to make API calls right away. The access token generated will be assigned your application's current permission level.

Token Actions

Create my access token

Figure 8-16. Generate access tokens

Your Access Token

This access token can be used to make API requests on your own account's behalf. Do not share your access token secret with anyone.

Access Token	
Access Token Secret	
Access Level	Read and write
Owner	joshuacook
Owner ID	

Figure 8-17. *Access token and access token secret*

Once these items have been obtained, enter the values into a cell in a new Jupyter Notebook as strings, as in Listing 8-40. Note that the code pattern has been included for easy copy-and-paste, but readers will need to replace each None value with the appropriate credential value as a string.

Listing 8-40. Load Twitter Credentials as Strings

```
In [9]: CONSUMER_KEY = None
        CONSUMER_SECRET = None
        ACCESS_TOKEN = None
        ACCESS_SECRET = None
```

In Listing 8-41, you install the twitter library using a system call, as you have done previously with pymongo and redis.

Listing 8-41. Install the twitter Library

```
In [10]: !pip install twitter
         Collecting twitter
           Downloading twitter-1.17.1-py2.py3-none-any.whl (55kB)
              100% |████████████████████████████| 61kB 932kB/s ta 0:00:011
         Installing collected packages: twitter
         Successfully installed twitter-1.17.1
```

You next instantiate a twitter.OAuth object using the Python twitter module and the credentials you have just loaded (Listing 8-42). You will use this object to facilitate your connection to Twitter's API.

Listing 8-42. Instatiate the twitter.OAuth Object

```
In [11]: from twitter import OAuth
         oauth = OAuth(ACCESS_TOKEN, ACCESS_SECRET, CONSUMER_KEY, CONSUMER_SECRET)
```

Collect Tweets by Geolocation

For this example, you will be using Twitter's Public Stream.[12] Applications that are able to connect to a streaming endpoint will receive a sample of public data flowing through Twitter and will be able to do so without polling or concern of API rate limits. In other words, the Public Stream is a safe and sanctioned way to collect a sample of live public tweets. That said, even this sample will return a great deal of unordered data.

In order to provide a modicum of order to your Twitter stream, you will restrict incoming tweets using a geolocation[13] bounding box,[14] or bbox. You can easily obtain a bbox for a location of interest using the Klokantech BoundingBox Tool.[15] In Figure 8-18, you obtain a bbox for Santa Monica, California in the United States, making sure to select CSV Raw as the copy and paste format.

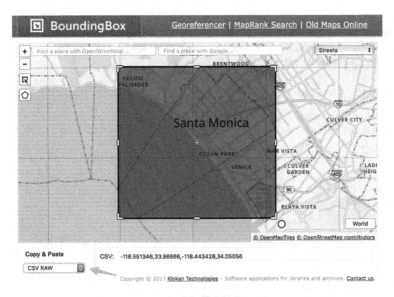

Figure 8-18. *Obtain a bbox for Santa Monica, California*

When you use a bbox to filter tweets, you will obtain only geolocated tweets falling within the bbox. Each bounding box should be specified as a pair of longitude and latitude pairs, with the southwest corner of the bounding box coming first, following geoJSON order (longitude, latitude).

Having obtained these values, you use them to define your bbox (Listing 8-43), defining the CSV list of values as a string.

---

[12]https://dev.twitter.com/streaming/public
[13]https://dev.twitter.com/streaming/overview/request-parameters#locations
[14]http://wiki.openstreetmap.org/wiki/Bounding_Box
[15]http://boundingbox.klokantech.com

Listing 8-43. Define a bbox for Santa Monica, California

```
In [12]: los_angeles_bbox = "-118.551346,33.96666,-118.443428,34.05056"
```

Finally, you instantiate a `twitter.TwitterStream`[16] object you will use to collect tweets (Listing 8-44). `twitter.TwitterStream` provides an interface to the Twitter Stream API in Python. The result of calling a method on this object is an iterator[17] that yields tweets decoded from the Twitter stream as JSON objects.

Listing 8-44. Instatiate a TwitterStream and an Associated Iterator

```
In [13]: from twitter import TwitterStream

         twitter_stream = TwitterStream(auth=oauth)
         twitterator = twitter_stream.statuses.filter(locations=los_angeles_bbox)
```

Following the instantiation of your `twitterator`, each subsequent next call will yield a tweet (Listing 8-45), a massive, nested JSON object.

Listing 8-45. Obtain the Next Tweet

```
In [14]: next(twitterator)
         {
           ...
           'created_at': 'Sun May 28 20:20:10 +0000 2017',
           ...
           'place': {'attributes': {},
            'bounding_box': {'coordinates': [[[-118.668404, 33.704538],
               [-118.668404, 34.337041],
               [-118.155409, 34.337041],
               [-118.155409, 33.704538]]],
             'type': 'Polygon'},
           ...
           },
           ...
           'quoted_status': {
           ...
            'text': "I'm in LA now and it's freakin' awesome 🌴",
           ...
           }
         }
```

---

[16]https://github.com/sixohsix/twitter#the-twitterstream-class
[17]https://docs.python.org/3/howto/functional.html?#functional-howto-iterators

Insert Tweets Into Mongo

Twitter is a wonderful source of messy, "real" data. Wrangling it into a database is where MongoDB truly shines. Using your `twitterator` object and the `.insert_one()` class function this can be done in a single line of code (Listing 8-46).

Listing 8-46. Insert the Next Tweet into `my_collection`

```
In [14]: my_collection.insert_one(next(twitterator))
Out[14]: <pymongo.results.InsertOneResult at 0x7f697c03cbd0>
```

You can verify the tweet's insertion and retrieve it with the `.count()` (Listing 8-47) and `.find_one()` class functions (Listing 8-48).

Listing 8-47. Count Objects in `my_collection`

```
In [15]: my_collection.count()
Out[15]: 1
```

Listing 8-48. Read One Object from `my_collection`

```
In [15]: my_collection.find_one()
Out[16]: {'_id': ObjectId('592b57ae042cee05c85ecf1d'),
          ...
          'text': '@sccrphobia11 Cubs 4 Dodgers 9 All runs in game were
          scored by homers by both
          teams. 7 HRs in all through 7 innings.',
          ...
          }
```

PostgreSQL

PostgreSQL[18] is an open source, object-relational database system. I favor the use of PostgreSQL over other structured query language (SQL) databases because of its wide adoption by the industry as well as for its native array and binary object types. A major selling point of SQL databases is their capacity for working with structured data. The creation of these structures is in and of itself an art and well beyond the scope of this text. Herein, I will emphasize PostgreSQL's natural aptitude for working with CSV files. I will also briefly explain PostgreSQL's array and binary types as natural partners to our work in the `numpy`/`scipy` stack.

You will continue to use the `docker volume` tool to persist data beyond the lifespan of a container. You will look at yet a third approach to managing networking in Docker, exploring the `docker network` tool.

---

[18]`www.postgresql.org/about/`

■ **Note** The configuration of network connections using `docker network` should be considered an advanced technique. You may skip this section with no peril to your learning and connect to PostgreSQL using either of the techniques outlined above.

If you do attempt to configure their connection to PostgreSQL using `docker network`, I assume that you are working on the host system where you have been running Jupyter.

Pull the `postgres` Image

As before, you retrieve the postgres image from its public page on Docker Hub.[19] In Listing 8-49, you pull the image.

Listing 8-49. Pull the postgres Image

```
$ docker pull postgres
Using default tag: latest
latest: Pulling from library/postgres
...
Digest: sha256:a2e6e6012a9056fa7647df5746119768bdb0bf4e82bb04819d5a8e450968a967
Status: Downloaded newer image for postgres:latest
```

Create New Data Volume

Following the previous pattern, you can think of PostgreSQL as a "service" comprised of the container running postgres and an associate data volume (Figure 8-19). In Listing 8-50, you create a new data volume.

Listing 8-50. Create a New postgres Data Volume

```
$ docker volume create --name pg-datastore
```

---

[19]https://hub.docker.com/_/postgres/

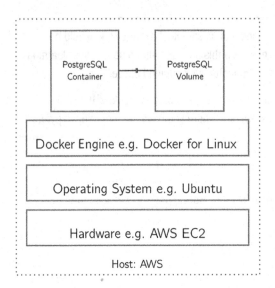

Figure 8-19. *PostgreSQL as a persistent service being mangaged by Docker*

Launch PostgreSQL as a Persistent Service

Again, you connect to the new data volume at runtime (Listing 8-51).

■ **Note**   You mount the Docker volume to the `/var/lib/postgresql/data` directory in the Postgres container.

Listing 8-51.   Display the Default Docker Networks

```
$ docker run -d --name this_postgres -v pg-dbstore:/var/lib/postgresql/data
-p 5432:5432 postgres
f9751d99f09d691dc3f04d246c502f1fa4ff3ae059632428a65affa3e5307f17
```

Verify PostgreSQL Installation

You can verify your PostgreSQL service by connecting to the running container via the PostgreSQL client, `psql`, issued via `docker exec` (Listing 8-53). Note that in order to do this you must specify a `DBNAME` and a `USERNAME` (Listing 8-52). As defined in the `postgres` image, the default value for both of these is `postgres`.

Listing 8-52. Display psql Help

```
$ docker exec this_postgres psql --help
psql is the PostgreSQL interactive terminal.

Usage:
  psql [OPTION]... [DBNAME [USERNAME]]
...
```

Listing 8-53. Connect to postgres, Create a Table, and Insert and Select a record

```
$ docker exec -it this_postgres psql postgres postgres
psql (9.6.3)
Type "help" for help.

postgres=# CREATE TABLE test ( _id INTEGER, name TEXT);
CREATE TABLE
postgres=# INSERT INTO test VALUES (1,'Joshua');
INSERT 0 1
postgres=# SELECT * FROM test LIMIT 1;
 _id |  name
-----+--------
   1 | Joshua
(1 row)

postgres=# \q
```

Docker Container Networking

You can connect two containers, such as a postgres service and a jupyter service, using Docker networking. By default, the Docker installation creates three networks for you. You will not need to create any additional networks; however, you will need to know how the default networks function. Listing 8-54 shows the three networks created for you by using the docker network ls command.

Listing 8-54. Display the Default Docker Networks

```
$ docker network ls
NETWORK ID          NAME                DRIVER
c9fced8bcbc9        bridge              bridge
3bec69979ce4        host                host
d8d9192909eb        none                null
```

In this text, you will only use the bridge network. It is the network that new containers connect to by default at runtime. It is possible to see this network on a Unix-variant host machine by running the ifconfig command, as seen in Listing 8-55.

Listing 8-55. Display the Host Network Configuration

```
$ ifconfig
...
docker0    Link encap:Ethernet  HWaddr 02:42:e4:f6:31:5a
           inet addr:172.17.0.1  Bcast:0.0.0.0  Mask:255.255.0.0
           inet6 addr: fe80::42:e4ff:fef6:315a/64 Scope:Link
           UP BROADCAST RUNNING MULTICAST  MTU:1500  Metric:1
           RX packets:49037 errors:0 dropped:0 overruns:0 frame:0
           TX packets:38968 errors:0 dropped:0 overruns:0 carrier:0
           collisions:0 txqueuelen:0
           RX bytes:31550630 (31.5 MB)  TX bytes:4105362 (4.1 MB)
...
```

You can connect to a running postgres container using a bash shell via the docker exec command, as seen in Listing 8-56.

Listing 8-56. Connect to this_postgres

```
$  docker exec -it this_postgres bash
root@f9751d99f09d:/#
```

From within the running container, you can see its network configuration by using the same ifconfig command (Listing 8-57). Note that you will first have to install the tool ifconfig using the package manager apt.

Listing 8-57. Container Network Configuration

```
root@f9751d99f09d:/data# apt update
Get:1 http://security.debian.org jessie/updates InRelease [63.1 kB]
...
Fetched 9949 kB in 3s (2535 kB/s)

root@f9751d99f09d:/data# apt install net-tools
...

root@f9751d99f09d:/data# ifconfig
...
eth0       Link encap:Ethernet  HWaddr 02:42:ac:11:00:04
           inet addr:172.17.0.2  Bcast:0.0.0.0  Mask:255.255.0.0
           inet6 addr: fe80::42:acff:fe11:4/64 Scope:Link
           UP BROADCAST RUNNING MULTICAST  MTU:1500  Metric:1
           RX packets:1715 errors:0 dropped:0 overruns:0 frame:0
           TX packets:492 errors:0 dropped:0 overruns:0 carrier:0
           collisions:0 txqueuelen:0
           RX bytes:10299480 (9.8 MiB)  TX bytes:35243 (34.4 KiB)
...

root@f9751d99f09d:/data# exit
```

You can see in Listing 8-57 that the Postgres container has an address of `172.17.0.2` on the network on which it is running. In Listing 8-55, you saw that the docker bridge network exists on `172.17.0.1` on your host machine.

■ **Note** This inspection of the network from within the container is not necessary for configuring the network. It is done solely for demonstration purposes.

Next, in Listing 8-58, you inspect the `bridge` network using the `docker network` inspect tool.

Listing 8-58. Inspect the Bridge Network

```
$ docker network inspect bridge
[
    {
        "Name": "bridge",
        "Id": "c9fced8bcbc9279ec29d880199e20795777c0bf9b2e7578f0d594a03981ff524",
        "Scope": "local",
        "Driver": "bridge",
        "EnableIPv6": false,
        "IPAM": {
            "Driver": "default",
            "Options": null,
            "Config": [
                {
                    "Subnet": "172.17.0.1/16"
                }
            ]
        },
        "Internal": false,
        "Containers": {
            "f9751d99f09d691dc3f04d246c502f1fa4ff3ae059632428a65affa3e5307f17": {
                "Name": "this_postgres",
                "EndpointID": "a9820c6df3120c7fc4a98f09372b1e51252
                               ceb937aaa18f7a9eec001cc6e2760",
                "MacAddress": "02:42:ac:11:00:04",
                "IPv4Address": "172.17.0.2/16",
                "IPv6Address": ""
            }
        },
    ...
    }
]
```

Here you can see the `bridge` network with address 172.17.0.1 and the Postgres container running on the network on 172.17.0.2.

Minimally Verify the Jupyter-PostgreSQL Connection

Next, you will attempt to connect to the running Postgres container using a Jupyter Notebook. In Listing 8-59, you launch a new Jupyter container using the docker run command.

Listing 8-59. Launch a Jupyter Container

```
$ docker run -v `pwd`:/home/jovyan -p 8888:8888 jupyter/scipy-notebook
```

First, you identify the running Jupyter container by using the docker ps command, as shown in Listing 8-60.

Listing 8-60. Show Running Containers

```
$ docker ps
CONTAINER ID IMAGE       COMMAND             CREATED        STATUS        PORTS
f9751d99f09d postgres    "docker-entrypoint.sh" 22 minutes ago  Up 22 minutes ...
cce1148863a2 jupyter/scipy-notebook
                         "tini -- start-notebo" 2 minutes ago   Up 2 minutes ...
```

As previously noted, when you launch a new container, by default it will be connected to the bridge network.

In Listing 8-61, you verify this using the docker network inspect tool.

Listing 8-61. Inspect Bridge Network

```
$ docker network inspect bridge
docker network inspect bridge
[
    {
        "Name": "bridge",
        ...
        "Containers": {
            "12fe7cea2e63b622c7804d1df96fbe2afce25d014e850b4fdec4e2e5498fde1b": {
                "Name": "this_postgres",
                "EndpointID": "a9820c6df3120c7fc4a98f09372b1e51252ceb937
                            aaa18f7a9eec001cc6e2760",
                "MacAddress": "02:42:ac:11:00:04",
                "IPv4Address": "172.17.0.2/16",
                "IPv6Address": ""
            },
            "cce1148863a22d11272ca031ded06139b2f0372d92aca269fd0d50234a30cf1c": {
                "Name": "hungry_cray",
                "EndpointID": "cde785070465ea79d4d9296895cb09f5975ec06a22
                            caca090fab789ca10b1d90",
                "MacAddress": "02:42:ac:11:00:03",
```

```
        "IPv4Address": "172.17.0.3/16",
        "IPv6Address": ""
    },
  },
 ...
]
```

Here you see that the Postgres container is still running on 172.17.0.2 and the jupyter/scipy-notebook container is running on 172.17.0.3. More importantly, because both of these containers are on the same network, you will be able to connect to one from the other. In Listing 8-62, you verify this by opening a bash shell to the jupyter/scipy-notebook container and verifying that you can ping the Postgres container at its IP on the bridge network.

Listing 8-62. Ping the Postgres Container from the jupyter/scipy-notebook Container

```
$ docker exec -it hungry_cray bash
jovyan@cce1148863a2:~$ ping -c 3 172.17.0.2
PING 172.17.0.2 (172.17.0.2): 56 data bytes
64 bytes from 172.17.0.2: icmp_seq=0 ttl=64 time=0.138 ms
64 bytes from 172.17.0.2: icmp_seq=1 ttl=64 time=0.103 ms
64 bytes from 172.17.0.2: icmp_seq=2 ttl=64 time=0.116 ms
--- 172.17.0.2 ping statistics ---
3 packets transmitted, 3 packets received, 0% packet loss
round-trip min/avg/max/stddev = 0.103/0.119/0.138/0.000 ms
```

Connnecting Containers by Name

To use the bridge network in order to connect to another Docker container, you are going to need to identify the IP address of that container on the bridge network, which, as you have just seen, a fairly non-trivial process. For your purposes, it may be easier to create an additional network and connect your containers to that network. If you have been following to this point, you should have two containers currently running, a jupyter/scipy-notebook and a Postgres container, as shown in Listing 8-63.

Listing 8-63. Show Running Containers

```
$ docker ps
CONTAINER ID IMAGE           COMMAND               CREATED         ... NAMES
f9751d99f09d postgres       "docker-entrypoint.sh" 22 minutes ago... this_postgres
cce1148863a2 jupyter/scipy-notebook
                            "tini -- start-notebo" 2 minutes ago... hungry_cray
```

Using the docker network tool, it will be possible to create a network and to connect these running containers to that network. Once you have connected these containers to the new network, you will be able to connect to them by name (by this_postgres and hungry_cray).

First, you create a new `bridge` network to be used by your containers (Listing 8-64) using the `docker network create` command.

Listing 8-64. Create a New `bridge` Network

```
$ docker network create jupyter_bridge
b8146c1af3a91abe4c123b9234d372e098fd71f1f0facd3e8251da2e864253ee
```

You can inspect the new network using the `docker network inspect` command, as shown in Listing 8-65.

Listing 8-65. Inspect the New `bridge` Network

```
$ docker network inspect jupyter_bridge
[
    {
        "Name": "jupyter_bridge",
        "Id": "b8146c1af3a91abe4c123b9234d372e098fd71f1f0facd3
            e8251da2e864253ee",
        "Scope": "local",
        "Driver": "bridge",
        "EnableIPv6": false,
        "IPAM": {
            "Driver": "default",
            "Options": {},
            "Config": [
                {
                    "Subnet": "172.18.0.0/16",
                    "Gateway": "172.18.0.1"                          }
            ]
        },
        "Internal": false,
        "Containers": {},
        "Options": {},
        "Labels": {}
    }
]
```

Here you see the new network you have created. You can see that it has no containers currently connected. In Listing 8-66, you connect the `jupyter/scipy-notebook` container, `hungry_cray`, and the Postgres container, `this_postgres`, to the new network you have created, after which you inspect the network once more.

Listing 8-66. Add the Running Containers to the New Bridge Network

```
$ docker network connect jupyter_bridge hungry_cray
$ docker network connect jupyter_bridge this_postgres
$ docker network inspect jupyter_bridge
[
```

```
{
    "Name": "jupyter_bridge",
    ...
    "Containers": {
        "cce1148863a22d11272ca031ded06139b2f0372d92aca269fd0d50234a30
        cf1c": {
            "Name": "hungry_cray",
            "EndpointID": "e01136684ea417200178a0afaaf634f39cd92f332
                          cb91b34a81dd7f7fcbbfc43",
            "MacAddress": "02:42:ac:19:00:02",
            "IPv4Address": "172.25.0.2/16",
            "IPv6Address": ""
        },
        "f9751d99f09d691dc3f04d246c502f1fa4ff3ae059632428a65affa3
        e5307f17": {
            "Name": "this_postgres",
            "EndpointID": "ae9a95c157092da3adbcb251ff9c370820f6b7ad9
                          f20a8df44cb971352af4cc9",
            "MacAddress": "02:42:ac:19:00:03",
            "IPv4Address": "172.25.0.3/16",
            "IPv6Address": ""
        }
    },
    ...
    }
]
```

Because you have connected these two containers to the same non-default bridge network, they will be able to resolve each other's location by container name, as shown in Listing 8-67.

Listing 8-67. Ping the Postgres Container from the jupyter/scipy-notebook Container

```
$ docker exec -it hungry_cray bash
jovyan@cce1148863a2:~$ ping -c 3 this_postgres
PING this_redis (172.25.0.3): 56 data bytes
64 bytes from 172.25.0.3: icmp_seq=0 ttl=64 time=0.248 ms
64 bytes from 172.25.0.3: icmp_seq=1 ttl=64 time=0.082 ms
64 bytes from 172.25.0.3: icmp_seq=2 ttl=64 time=0.081 ms
--- this_redis ping statistics ---
3 packets transmitted, 3 packets received, 0% packet loss
round-trip min/avg/max/stddev = 0.081/0.137/0.248/0.078 ms
```

■ **Note** Containers that are solely linked by the default bridge network will not be able to resolve each other's container name.

Using PostgreSQL with Jupyter

You will use the same established pattern to connect to PostgreSQL from within a Jupyter notebook. You will first perform a !pip install of the psycopg2 library You will use to connect to the PostgreSQL database (Listing 8-68).

Listing 8-68. Install psycopg2 via a Shell Call from a Jupyter Notebook

```
In [1]: !pip install psycopg2
        Collecting psycopg2
          Downloading psycopg2-2.7.1-cp35-cp35m-manylinux1_x86_64.whl (2.7MB)
            100% |████████████████████████████| 2.7MB 333kB/s eta 0:00:01
        Installing collected packages: psycopg2
        Successfully installed psycopg2-2.7.1
```

Jupyter, PostgreSQL, Pandas, and psycopg2

Pandas, as nearly every other technology referenced in this tome, is an open source library. pandas is *the* Python data analysis library. It plays well with the entire numerical Python stack from numpy to scikit-learn. It is intuitive and easy to use and has a place on the tool belt of every data scientist. Heck, it has even been known to make the R programmer more comfortable in Python land. Here, you will use pandas where appropriate to supplement psycopg2's connection to PostgreSQL.

Minimal Verification

You will start out by performing a minimal verification of your connection. You will

1. Import a few libraries for your work (Listing 8-69).

2. Instantiate a connection and a cursor attached to that connection (Listing 8-70).

3. Use the cursor to execute a query (Listing 8-71).

4. Display the results of the query in a pandas.DataFrame (Listing 8-72).

5. Close the connection (Listing 8-73).

■ **Note**   You import an additional module in Listing 8-69, psycopg2.extras. You will use this module in Listing 8-70 to specify that you wish to use a special cursor type, the psycopg2.extras.RealDictCursor. Using this cursor type will allow you to easily pass the results of your query to a pandas.DataFrame, passing column names seamlessly from the database to the data frame.

Listing 8-69. Import Necessary Libraries

```
In [2]: import pandas as pd
        import psycopg2 as pg2
        import psycopg2.extras as pgex
```

Listing 8-70. Instantiate Connection and Cursor

```
In [3]: con = pg2.connect(host='this_postgres', user='postgres',
database='postgres')
        cur = con.cursor(cursor_factory=pgex.RealDictCursor)
```

Listing 8-71. Execute Query

```
In [4]: cur.execute("SELECT * FROM test;")
```

Listing 8-72. Fetch All Results from the Query and Display in a pandas.DataFrame

```
In [5]: pd.DataFrame(cur.fetchall())
Out[5]:        _id      name
        0       1       Joshua
```

Listing 8-73. Close the Connection

```
In [6]: con.close()
```

Loading Data into PostgreSQL

Loading data into PostgreSQL can be challenging. In later chapters, you will explore the use of Dockerfiles to load data and hold that this is the best practice. For the quick insertion of data, such as is necessary for exploring some minimal Jupyter-Postgres interaction in the remainder of this chapter, you will use simple SQL statements from Jupyter and manage your transactions manually via BEGIN and COMMIT.

In Listing 8-74, you create a new table called from_jupyter_test with four data types: INTEGER, TEXT, DOUBLE PRECISION[], BYTEA. While readers are no doubt familiar with the first two data types, at this time I call special attention to the second two. The trailing brackets on the DOUBLE PRECISION[] type signifies that this is an array of DOUBLE PRECISION (floating-point) values. BYTEA is a binary type and you will use it to hold numpy arrays much as you did earlier with Redis.

Listing 8-74. Create a Table in PostgreSQL from Juptyer

```
In [7]: con = pg2.connect(host='this_postgres', user='postgres',
database='postgres')
        cur = con.cursor(cursor_factory=pgex.RealDictCursor)
        cur.execute("""
        BEGIN;
        CREATE TABLE from_jupyter_test (
            _id INTEGER,
```

```
        name TEXT,
        list DOUBLE PRECISION[],
        vector BYTEA
    );
    COMMIT;
    """)
```

In Listing 8-75, you insert two rows into the database. Note the special handling of a list type on insertion. Note also that you have only inserted three values. By default these will align with the first three columns in the table.

Listing 8-75. Insert Two Rows into PostgreSQL

```
In [8]: cur.execute("""
        BEGIN;
        INSERT INTO from_jupyter_test VALUES (1, 'spam', '{1,2,3,4,5}');
        INSERT INTO from_jupyter_test VALUES (2, 'eggs', '{1,4,9,16,25}');
        COMMIT;
        """)
```

Finally, you query the from_jupyter_test table and display the results in a pandas.DataFrame (Listing 8-76), before closing the connection (Listing 8-77).

Listing 8-76. Select All Rows from_jupyter_test and Display in a pandas.DataFrame

```
In [9]: cur.execute("""
        SELECT * FROM from_jupyter_test;""")
        pd.DataFrame(cur.fetchall())
Out[9]:          _id   list                          name    vector
        0         1     [1.0, 2.0, 3.0, 4.0, 5.0]     spam    None
        1         2     [1.0, 4.0, 9.0, 16.0, 25.0]   eggs    None
```

Listing 8-77. Close the Connection

```
In [10]: con.close()
```

PostgreSQL Binary Type and Numpy

Like Redis, PostgreSQL has a native binary type, BYTEA, making it ideal for computational work with numpy. Converting to and from this type is a reasonably straightforward process. Here, you

1. Query all lists in your PostgreSQL database (Listing 8-78) and display them as native Python objects.

2. Convert these lists to numpy.array objects (Listing 8-79).

3. Convert the numpy.array objects to PostgreSQL binary objects using psycopg2.Binary (Listing 8-80).

4. Perform two SQL updates on your database, setting the vector value in the table to be the respective binary object (Listing 8-81).

5. Query all values and display in a pandas.DataFrame (Listing 8-82).

6. Query the vector values, fetch them one at a time, and convert them back to numpy.array objects (Listing 8-83).

Listing 8-78. Query List Values and Display Results

```
In [11]: cur.execute("""
         SELECT list FROM from_jupyter_test;""")
         results = cur.fetchall()
         results
Out[11]: [{'list': [1.0, 2.0, 3.0, 4.0, 5.0]}, {'list': [1.0, 4.0, 9.0, 16.0, 25.0]}]
```

Listing 8-79. Convert Lists to numpy.array Objects

```
In [12]: import numpy as np
         ary_1 = np.array(results[0]['list'])
         ary_2 = np.array(results[1]['list'])
```

Listing 8-80. Convert numpy.array Objects to Binary Objects

```
In [13]: bin_ary_1 = pg2.Binary(ary_1)
         bin_ary_2 = pg2.Binary(ary_2)
```

Listing 8-81. Perform SQL Updates

```
In [14]: update_sql = """
         BEGIN;
         UPDATE from_jupyter_test
         SET vector = {}
         WHERE _id={};
         COMMIT;
         """.format(pg2.Binary(bin_ary_1), 1)
         cur.execute(update_sql)

In [15]: update_sql = """
         BEGIN;
         UPDATE from_jupyter_test
         SET vector = {}
         WHERE _id={};
         COMMIT;
         """.format(pg2.Binary(bin_ary_2), 2)
         cur.execute(update_sql)
```

Listing 8-82. Query and Display All Values

```
In [16]: cur.execute("""
         SELECT * FROM from_jupyter_test;""")
         pd.DataFrame(cur.fetchall())
Out[16]:    _id   list                          name   vector
    0        1    [1.0, 2.0, 3.0, 4.0, 5.0]     spam   [b'\x00', b'\x00', ...
    1        2    [1.0, 4.0, 9.0, 16.0, 25.0]   eggs   [b'\x00', b'\x00', ...
```

Listing 8-83. Convert Vector Value to numpy.array

```
In [17]: cur.execute("""
         SELECT vector FROM from_jupyter_test;""")
         result = cur.fetchone()
         result
Out[17]: {'vector': <memory at 0x7f8fd643b708>}

In [18]: np.frombuffer(result['vector'])
Out[18]: array([ 1.,   2.,   3.,   4.,   5.])

In [19]: result = cur.fetchone()

In [20]: np.frombuffer(result['vector'])
Out[20]: array([ 1.,    4.,    9.,   16.,   25.])
```

Summary

This is one of the longest chapters in the book and represents a significant departure from the material to this point. I introduced three data stores (Redis, MongoDB, and PostgreSQL) and discussed how to use them with Docker, especially with regard to data persistence and networking. I discussed ways in which numpy vectors can be serialized for storage in both Redis and PostgreSQL. I introduced Docker Volumes and Docker Networks.

This is a chapter that bears repeated readings and that I hope will serve as a reference for future projects. I hope that you complete the chapter with an understanding of when each of the three data stores might be used, and I hope you're confident that you will be able to configure any one of them using this chapter as reference.

We are drawing near to the end of the text. In the last two chapters, we will make extensive use of these data stores, looking at how we can use Docker Compose to build larger applications and finally revisiting the idea of interactive programming and what it might look like to build software under this paradigm.

CHAPTER 9

■ ■ ■

Docker Compose

Thus far, I have focused the discussion on single containers or individually managed pairs of containers running on the same system. In this chapter, you'll extend your ability to develop applications comprised of multiple containers using the Docker Compose tool.

The provisioning of systems refers to installation of necessary libraries, the configuration of resource allocation and network connections, and the management of persistent and state-specific data. In the past, the provisioning and maintenance of a system might have been done manually by a system administrator or operations engineer. In recent years, the DevOps community has thrived around a concept for building software applications called "infrastructure as code" (IaC). IaC is the practice of using software developed for the specific purpose of provisioning systems, as opposed to doing this manually.

You have already seen how the Docker toolset can be used to provision systems. Using a Dockerfile, it is possible to use code to provision the system libraries and Python libraries required by a specific container. In this chapter, you will explore how to use the Docker Compose tool to configure resource allocation, connections between multiple containers, environment variables, and other state-specific data, in addition to persistent data. Having mastered these concepts, you will begin to think about the multi-container systems you design as single, containerized applications. Furthermore, using Docker Compose you will be able to start, stop, and scale your applications using a simple command line interface.

As you proceed, I will discuss what all of this means to the data scientist. Specifically, I will emphasize three powerful advantages to working in this way, namely the ability to

1. Develop locally and deploy anywhere

2. Quickly and easily share complex applications with stakeholders

3. Personally manage infrastructure without needing to rely on IT resources

Install docker-compose

If you're running Docker for Linux, docker-compose can be installed using the instructions provided here: https://github.com/docker/compose/releases. This will include those of you who have been following along using the recommended best practice of installing on a disposable cloud-based system. Those using Docker for

Linux will want to follow these instructions. If you're using Docker for Mac, Docker for Windows, or Docker Toolbox, you may skip this next section, as docker-compose is installed by default upon Docker installation.

As of the writing of this text, the process is as outlined. The first set of commands (Listing 9-1) retrieves the latest compiled binary from GitHub[1] and then moves the binary to an appropriate location on your system. The second command (Listing 9-2) allows the binary to be executed by the local system.

Listing 9-1. Download the docker-compose Binary

```
$ sudo curl -L https://github.com/docker/compose/releases/download/1.15.0/
docker-compose-`uname -s`-`uname -m` > docker-compose
sudo mv docker-compose /usr/local/bin/docker-compose
  % Total    % Received % Xferd  Average Speed   Time    Time     Time  Current
                                 Dload  Upload   Total   Spent    Left  Speed
100   600    0   600    0     0   2327      0 --:--:-- --:--:-- --:--:--  2334
100 8066k 100 8066k    0     0  1122k      0  0:00:07  0:00:07 --:--:-- 1556k
```

Listing 9-2. Grant Execute Permissions to the docker-compose Binary

```
$ sudo chmod +x /usr/local/bin/docker-compose
```

In Listing 9-3, you test your docker-compose installation.

Listing 9-3. Test docker-compose

```
$ docker-compose --version
docker-compose version 1.12.0, build b31ff33
```

What Is docker-compose?

Docker Compose is a tool for managing a multi-container application with Docker. An application is defined by Compose using a single docker-compose.yml file. The developer maintains a directory of source code defining an application. This directory may include a library of Python or C files; zero or more Dockerfiles defining containers; raw data files such as CSVs, feather,[2] or JSON files; and other provisioning scripts such as a requirements.txt file for a Python project. The docker-compose.yml file sits at the top level of this directory. The application is built, run, stopped, and removed using a single docker-compose command. The docker-compose tool refers to the file as it communicates with the Docker engine, as if the entire application were a single Docker container.

---

[1]https://github.com/docker/compose
[2]https://blog.rstudio.org/2016/03/29/feather/

Docker Compose Versions

The Docker Compose file syntax is currently on Version 3. For new projects, Docker recommends using the most recent version. That said, all versions are backward-compatible and many docker-compose.yml files found on the Internet use earlier versions. A version is specified as the first value in a docker-compose.yml file, as shown in Listing 9-4. If no version is specified, Version 1 will be used.

Listing 9-4. A docker-compose.yml File Showing a Version Specification

```
version: '3'
services:
  db:
    image: postgres
    volumes:
      - data:/var/lib/postgresql/data
  volumes:
    data:
      driver: mydriver
```

Build a Simple Docker Compose Application

You will build your first Docker Compose application to be a Jupyter Notebook Server running in conjunction with a Redis server (Figure 9-1).

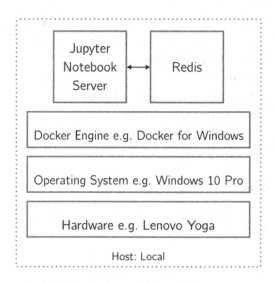

Figure 9-1. *A simple Docker Compose application*

You will first create a directory to hold your project (Listing 9-5).

Listing 9-5. Create a Directory to Hold the Project

```
$ mkdir ch_9_jupyter_redis
$ cd ch_9_jupyter_redis
```

Next (Listing 9-6), you will use the command line text editor vim[3] to create a docker-compose.yml file (Listing 9-7) to define this application.

Listing 9-6. Create docker-compose.yml

```
$ vim docker-compose.yml
```

Listing 9-7. jupyter_redis docker-compose.yml File

```
version: '3'
services:
  this_jupyter:
    image: jupyter/scipy-notebook
    ports:
    - 8888:8888
    volumes:
    - .:/home/jovyan
  this_redis:
    image: redis:alpine
```

You use the Compose file to define two services, this_jupyter and this_redis. The this_jupyter service

- Uses the jupyter/scipy-notebook from the Docker Hub registry, as specified by the image: keyword.

- Attaches the local directory (.) to the (hopefully familiar) jupyter WORKDIR, /home/jovyan,[4] as specified by the volumes: keyword.

- Forwards the exposed port 8888 to the port 8888 on the host machine, as specified by the ports: keyword.

- Links the container to the Redis container, as specified by the links: argument.

The redis service uses the redis image with tag alpine[5] from the Docker Hub registry.

---

[3]www.vim.org
[4]https://github.com/jupyter/docker-stacks/blob/master/base-notebook/Dockerfile#L80
[5]https://alpinelinux.org

■ **Note** The definition of every container defined in a docker-compose.yml file must begin with either the image: argument or the build: argument. I will discuss the build: argument in the next section of this chapter.

Run Your Application with Compose

You used the docker-compose.yml file to define your application. Now (Listing 9-8), you use the docker-compose command line tool to start the application. You will use the -d argument to specify that you wish to launch the application in detached mode.

Listing 9-8. Start the Compose Application jupyter_redis

```
$ docker-compose up
Creating network "ch9jupyterredis_default" with the default driver
Creating ch9jupyterredis_this_redis_1
Creating ch9jupyterredis_this_jupyter_1
```

The docker-compose up command first instructs the Docker Engine to either

1. Check the local image cache for the specified image if the container definition begins with the image: keyword.

2. Build the image from a referenced Dockerfile if the container definition begins with the build: keyword.

Here, you have passed an image: keyword for both containers. In this case, both images with which you are working are currently in your image cache. Were they not in the cache, the Docker engine would pull them from Docker Hub.

Next, the docker-compose up command instructs the Docker engine to create a network over which the application's containers can communicate. Here, a network called ch9jupyterredis_default was created. Listing 9-14 shows how to use this network.

Finally, the docker-compose up command instructs the Docker engine to create the containers defined in the docker-compose.yml file. This is equivalent to running the two commands in Listing 9-9, with one exception: the connection created by docker-compose is superior to the link created by a --link flag, as you will see in a moment.

■ **Warning** Don't execute Listing 9-9. It is included here to show what you *would* run were you to instantiate these two containers manually.

Listing 9-9. Manually Instantiate Two Containers as Defined in docker-compose.yml

```
$ docker run --name ch9jupyterredis_this_redis_1 redis
$ docker run -v `pwd`:/home/jovyan -p 8888:8888 --name ch9jupyterredis_this_
jupyter_1 --link ch9jupyterredis_this_redis_1 jupyter/scipy-notebook
```

This method is a mouthful and I hope that readers can see the superficial benefits of using docker-compose immediately in terms of making the definition of command line options much more straightforward.

In Listing 9-10, you use docker-compose ps to display process information for the containers associated with the current docker-compose.yml file.

Listing 9-10. Display Containers for Current docker-compose.yml

```
$ docker-compose ps
Name                                  Command
----------------------------------------------------------------------------
ch9jupyterredis_this_jupyter_1     tini -- start-notebook.sh
ch9jupyterredis_this_redis_1       docker-entrypoint.sh redis ...

State      Ports
----------------------------------------------------------------------------
Up         0.0.0.0:8888->8888/tcp
Up         6379/tcp
```

This command will not show information for other containers as docker ps will. In Listing 9-11, you change directories to go up a level. When you have done that, you have no docker-compose.yml file that can be referenced by the docker-compose command. When you request docker-compose ps information in a directory with no docker-compose.yml file, you receive an error. Following receiving the error, you return to your project directory.

Listing 9-11. Request docker-compose ps Information in a Directory with No docker-compose.yml

```
$ cd ..
$ ls
ch_9_jupyter_redis
$ docker-compose ps
ERROR:
        Can't find a suitable configuration file in this directory or any
        parent. Are you in the right directory?

        Supported filenames: docker-compose.yml, docker-compose.yaml
$ cd ch_9_jupyter_redis
```

You can operate on the containers you have created as you would any other Docker container. In Listing 9-12, you use docker exec to connect to your Jupyter container. From there you check env and pipe (|) this to a grep for the term redis, so that you can examine the environment variables associated with the Jupyter container's link to the Redis container.

Listing 9-12. Examine jupyter Environment Variables for redis

```
$ docker exec ch9jupyterredis_this_jupyter_1 bash
jovyan@63f28e183a88:~$ env | grep redis
```

Note that nothing is displayed. In Listing 9-13, you verify that this is true by displaying the full environment.

Listing 9-13. Examine jupyter Environment Variables

```
jovyan@db36cb53ea93:~$ env
HOSTNAME=db36cb53ea93
NB_USER=jovyan
SHELL=/bin/bash
TERM=xterm
LC_ALL=en_US.UTF-8
LS_COLORS= ...
PWD=/home/jovyan
LANG=en_US.UTF-8
SHLVL=1
HOME=/home/jovyan
LANGUAGE=en_US.UTF-8
XDG_CACHE_HOME=/home/jovyan/.cache/
DEBIAN_FRONTEND=noninteractive
CONDA_DIR=/opt/conda
NB_UID=1000
_=/usr/bin/env
```

In Chapter 8, Listing 8-12, you did much the same thing. You created a link to a running Redis container, although in that case you did so using the --link flag passed to the docker run argument. You then connected to the running Jupyter container via a bash shell and grepped the environment variables for THIS_REDIS. If you recall, you found several variables that could be used in place of referring to the Redis container by its local IP address. As can be seen, this is not the case when you created an application comprised of Jupyter and Redis via docker-compose. You are going to need to connect to Redis in a different way. Fortunately, as you will see in a moment, it is much, *much* easier.

While you're here, grab the security token for your Jupyter container so that you will be able to access Jupyter through your browser (Listing 9-14).

Listing 9-14. Obtain the jupyter Security Token via bash Shell Connected to the Container

```
jovyan@63f28e183a88:~$ jupyter notebook list
Currently running servers:
http://localhost:8888/?token=bca81e140ba43b4a3b20591812a6af32289fc66131e8e
5e0 :: /home/jovyan
```

As before, you will use this to access Jupyter in your browser, substituting localhost for the appropriate IP address, if necessary. In Figure 9-2, you navigate to your Jupyter site on a remote AWS t2.micro. You can see that the -v flag has made the lone file in your local directory, the docker-compose.yml file, available to the Jupyter Notebook server.

Figure 9-2. *Jupyter running on a t2.micro.*

Listing 9-15 shows a trivial connection to the Redis container from a Jupyter Notebook. Here, you make use of the link created by docker-compose. The link quite simply is the name of the service defined in the docker-compose.yml file, namely, this_redis.

Listing 9-15. Connect to redis from jupyter

```
In [1]: !pip install redis
        Collecting redis
          Downloading redis-2.10.5-py2.py3-none-any.whl (60kB)
            100% |████████████████████████████████| 61kB 2.7MB/s ta 0:00:011
        Installing collected packages: redis
        Successfully installed redis-2.10.5
        You are using pip version 8.1.2, however version 9.0.1 is available.
        You should consider upgrading via the 'pip install --upgrade pip'
        command.
In [2]: import redis
In [3]: REDIS = redis.Redis(host='this_redis')
In [4]: REDIS.incr('my_incrementor')
Out[4]: 1
In [5]: REDIS.get('my_incrementor')
Out[5]: b'1'
```

Finally, in Listing 9-16, you tear your simple application down using the docker-compose down command.

Listing 9-16. Tear Down Application via docker-compose down

```
$ docker-compose down
Stopping ch9jupyterredis_this_jupyter_1 ... done
Stopping ch9jupyterredis_this_redis_1 ... done
Removing ch9jupyterredis_this_jupyter_1 ... done
Removing ch9jupyterredis_this_redis_1 ... done
Removing network ch9jupyterredis_default
```

Note that not only does docker-compose down stop your containers, it removes them, before removing the customer network defined to connect your containers.

Jupyter and Mongo with Persistence

In Chapter 8, you configured a two-container system consisting of a Jupyter container and a MongoDB container running on two separate host systems on AWS. Here, you create the functionally equivalent system using Docker Compose. One advantage of using Docker Compose to build the system is that you will be able to run both services from the same host system without putting significant effort into managing your network configuration. You saw previously when configuring your simple Compose application that you were able to access Redis simply by using the name you had given to the service in your docker-compose.yml file. You will take advantage of this simple method for configuring networks and use two new Docker Compose techniques available: the configuration of data volumes and the definition of environment variables. Figure 9-3 shows a diagram of the system you will be configuring.

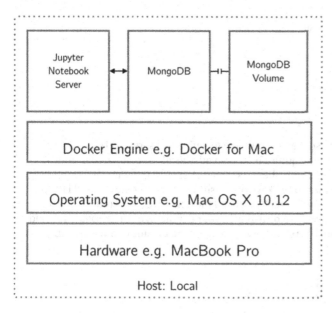

Figure 9-3. *A Docker Compose application with two service and a data volume*

Once more, you begin by creating a directory to hold your project (Listing 9-17). Recall that docker-compose uses the docker-compose.yml file to communicate with the Docker engine, and as such it is a best practice to create a new directory for each project and give it its own docker-compose.yml file. In Listing 9-18, you create the new docker-compose.yml file (Listing 9-19).

Listing 9-17. Create a Directory to Hold the Project

```
$ mkdir ch_9_jupyter_mongo
$ cd ch_9_jupyter_mongo
```

Listing 9-18. Create the docker-compose.yml file

```
$ vim docker-compose.yml
```

Listing 9-19. ch_9_jupyter_mongo/docker-compose.yml file

```
version: '3'
services:
  this_jupyter:
    build: docker/jupyter
    ports:
      - "8888:8888"
    volumes:
      - .:/home/jovyan
    env_file:
      - config/jupyter.env
this_mongo:
    image: mongo
    volumes:
      - mongo_data:/data/db
volumes:
  mongo_data:
```

While you once more use the Compose file to define two services, this_jupyter and this_mongo, you have done quite a bit more with this file as compared to your previous application. To begin with, the this_jupyter service is defined not by an image: keyword, but rather using the build: keyword. Listing 9-22 shows the Dockerfile you will be using for your build. Additionally, you add the env_file: keyword. Listing 9-24 shows the environment file you will be using. In your definition of this_mongo, you add a volumes: keyword that makes reference to a mongo_data volume defined later in the Compose file.

Specifying the Build Context

Here, you have specified the build keyword by providing a string that is a path to your desired build context. In Chapter 5, you noted that the build context refers to the collection of files that will be used to build the specific image. In other words, a build context is a directory that contains the Dockerfile to be used as well as any other files required by the build. Here, you specify a build context of docker/jupyter. In Listing 9-20, you create this directory relative to the location of your docker-compose.yml file. Note that the -p flag passed to the mkdir command allows you to create the nested directory structure. In this nested directory structure, you are using a best practice that will be discussed in Chapter 10.

Listing 9-20. Create this_jupyter Build Context

```
$ mkdir -p docker/jupyter
$ tree
.
├──── docker
│   └──── jupyter
└──── docker-compose.yml
```

Next (Listing 9-21), you are going to need to create the Dockerfile (Listing 9-22) that will define your new image. In defining your Dockerfile you are using the best practices defined in Chapter 7.

Listing 9-21. Create docker/jupyter/Dockerfile

```
$ vim docker/jupyter/Dockerfile
```

Listing 9-22. docker/jupyter/Dockerfile

```
FROM jupyter/scipy-notebook
USER root
RUN conda install --yes --name root spacy pymongo
RUN ["bash", "-c", "source activate root && pip install twitter"]
RUN python -m spacy download en
USER jovyan
```

■ **Note**   Since you don't intend on running any code using the Python 2 kernel, it is not necessary to install the libraries in the python2 environment. Instead, you only install to the root environment which, as you recall from Chapter 6, is the Python 3 environment.

Specify the Environment File

In Chapter 8, you obtained a set of OAuth credentials that could be used to stream tweets directly from Twitter's Streaming API. It is a best practice in terms of security to store these credentials in a separate file. Here, you will store those credentials in an environment file called jupyter.env. You have told Docker Compose to use this file using the env_file: keyword. In Listing 9-23, you create the config directory to hold this file and then create the new file shown in Listing 9-24. As before, replace the dummy strings in the environment file with your actual API credentials.

Listing 9-23. Create the config Directory and config/jupyter.env

```
$ mkdir config
$ vim config/jupyter.env
```

Listing 9-24. `config/jupyter.env`

```
CONSUMER_KEY=dummy_consumer_key
CONSUMER_SECRET=dummy_consumer_secrete
ACCESS_TOKEN=dummy_access_token
ACCESS_SECRET=dummy_access_secret
```

■ **Warning**   The `config/jupyter.env` file should be treated as a `bash` script. This means that the variable definition requires no spaces on either side of the equal sign. In `bash`, `var=1` is a variable assignment,[6] while `var = 1` is a Boolean comparison that will first try to execute `var1`.[7]

In Listing 9-25, you use the `tree` tool to share the final status of your application directory.

Listing 9-25.   Use `tree` to Show the Application Directory

```
$ tree
.
├──── config
│    └──── jupyter.env
├──── docker
│    └──── jupyter
│         └──── Dockerfile
└──── docker-compose.yml
```

Data Persistence

In Chapter 8, you used Docker data volumes to persist data beyond the lifespan of a container. You did this using the `docker  volume` tool, which in the context of infrastructure as code, you should think of as a manual method. Using a `docker-compose.yml` file, it is possible to define a volume in much the same way that a service is defined (that is, to specify the creation of a volume using code). Furthermore, you can define how the volume will be used by the application (that is, you can specify an attachment to a specific service). In Listing 9-19, you define a single volume called `mongo_ data` and then link that volume to your `mongo` service.

---

[6]`http://tldp.org/LDP/abs/html/varassignment.html`
[7]`http://tldp.org/LDP/abs/html/gotchas.html#WSBAD`

Build Your Application with Compose

Before running your application, you will need to build it. This is because at least one of your defined services uses the build: keyword as opposed to be image: keyword in order to define the image that will be used to instantiate its container. In Listing 9-26, you perform this build. It is worth noting that this command will only affect those services that are defined using the build: keyword. You can see that the build process skips this_mongo because it uses an image.

Listing 9-26. Build the Application Using docker-compose build

```
$ docker-compose build
this_mongo uses an image, skipping
Building this_jupyter
Step 1/6 : FROM jupyter/scipy-notebook
 ---> 3dc12029099d
Step 2/6 : USER root
 ---> Using cache
 ---> d0fe6db71e0b
Step 3/6 : RUN conda install --yes --name root spacy pymongo
 ---> Running in 26e516e316c3
...
Step 4/6 : RUN bash -c source activate root && pip install twitter
 ---> Running in 5a07f7033056
...
Step 5/6 : RUN python -m spacy download en
 ---> Running in 319588ae94c6
...
Step 6/6 : USER jovyan
 ---> Running in 5c0f78fd96f2
 ---> d928c6dcf4fd
Removing intermediate container 5c0f78fd96f2
Successfully built d928c6dcf4fd
Successfully tagged ch9jupytermongo_this_jupyter:latest
```

■ **Warning**   When docker-compose up is run, the Docker client checks the local image cache to see if the images associated with each defined service are present in the cache. For a service defined using the build: keyword, if the image is not in the cache, then the image will be built. This is to say that docker-compose build will be implicitly called. This will only happen if the image is not in the local image cache.

The implications of this are that if changes have been made to a Dockerfile or other application files, the docker-compose up command has no mechanism for picking up on the changes and triggering a build. To the uninitiated, it may seem as though a build is implicitly called by docker-compose up, but truthfully this only happens if the image is not in the cache. It is for this reason that it is a recommended best practice to always run docker-compose build before running docker-compose up.

Finally, having completed your build, you run your application, again using the docker-compose up command (Listing 9-27).

Listing 9-27. Start the Compose Application jupyter_mongo

```
$ docker-compose up -d
Starting ch9jupytermongo_this_mongo_1
Starting ch9jupytermongo_this_jupyter_1
ubuntu@LOCAL:~/ch_9_jupyter_mongo
```

In Listing 9-28, you again use the docker-compose ps tool to display process information for the containers associated with your current application.

Listing 9-28. Display Containers for Current docker-compose.yml

```
$ docker-compose ps
Name                                  Command
--------------------------------------------------------------
ch9jupytermongo_this_jupyter_1 tini -- start-notebook.sh
ch9jupytermongo_this_mongo_1    docker-entrypoint.sh mongod

State      Ports
--------------------------------------------------------------
Up         0.0.0.0:8888->8888/tcp
Up         27017/tcp
```

Next (Listing 9-29), you obtain the security token for your Jupyter container so that you will be able to access Jupyter through your browser.

Listing 9-29. Obtain jupyter Security Token via a Shell Call to the Container

```
$ docker exec ch9jupytermongo_this_jupyter_1 jupyter notebook list
Currently running servers:
http://localhost:8888/?token=0029b465c514ce18856a5a2751a95466504fac1
8b43531ce :: /home/jovyan
```

Finally, in Figure 9-4, you navigate to a browser and, using the IP associated with your host system, you access Jupyter.

Figure 9-4. Jupyter running on a t2.micro

In order to test your system, you will once more stream tweets using a location-based filter. As in Chapter 8, you will insert each tweet you collect into MongoDB. To add a level of complexity, prior to inserting the tweet into MongoDB, you will use the spaCy library to encode that tweet text as a numpy vector. In order to store the vector in MongoDB, you will need to serialize the vector as a binary bytestream. You did this with Redis in Chapter 8 and here you do the same with MongoDB.

You first configure your Twitter authentication. You import the environ object from the os module and the OAuth class from the twitter module. The environ object[8] is a mapping[9] object representing the operating system's string environment. It is captured at the time of import. Here, you will use it to reference to the environment variables containing your credentials as defined in the docker-compose.yml and config/jupyter.env files; see Listing 9-30.

Listing 9-30. Import Modules Necessary to Configure Twitter Authentication

```
In [1]: from os import environ
        from twitter import OAuth
```

In Listing 9-31, you instantiate an OAuth object using the stored credentials. Note that each value is accessed using its key similar to a dictionary.

Listing 9-31. Instantiate the OAuth Object

```
In [2]: oauth = OAuth(environ['ACCESS_TOKEN'],
                      environ['ACCESS_SECRET'],
                      environ['CONSUMER_KEY'],
                      environ['CONSUMER_SECRET'])
```

In Listing 9-32, you import the TwitterStream class and instantiate an object of that class using your defined authentication.

---

[8]https://docs.python.org/3/library/os.html#os.environ
[9]https://docs.python.org/3/glossary.html#term-mapping

Listing 9-32. Instantiate `TwitterStream`

```
In [3]: from twitter import TwitterStream

        los_angeles_bbox = "-118.55, 33.97, -118.44, 34.05"
        twitterator = (TwitterStream(auth=oauth)
                       .statuses
                       .filter(locations=los_angeles_bbox))
```

Finally, in Listing 9-33, you pull a single tweet from the stream and then display its keys.

Listing 9-33. Pull a Single Tweet and Display Its keys

```
In [4]: this_tweet = next(twitterator)
        this_tweet.keys()
```

You now have a tweet in memory and can insert it into MongoDB. Prior to insertion you will use the spaCy library to encode the tweet as a vector. You will add the encoded vector to the dictionary containing your tweet as a binary bytestream. First (Listing 9-34), you import spacy and load the en model. In doing so, on your t2.micro, you receive a memory error.

Listing 9-34. Import spacy and Load the en Language Model

```
In [5]: import spacy

        nlp = spacy.load('en')
        -------------------------------------------------------------------
        MemoryError                              Traceback (most recent call last)
        <ipython-input-8-e0448d429293> in <module>()
              1 import spacy
              2
        ----> 3 nlp = spacy.load('en')
```

Well, at first such an error might be annoying or even daunting, but being able to efficiently deal with issues like this is a primary reason why you are learning this technology in the first place. In Listing 9-35, you use the docker stats tool to diagnose your error. Note that docker-compose does not have its own docker stats tool so you simply use the standard tool you have been using previously.

Listing 9-35. Use docker stats to Diagnose a Memory Error

```
CONTAINER      CPU %    MEM USAGE / LIMIT      MEM %
698ba3322462   0.02%    539.2MiB / 990.7MiB    54.43%
a37b17785360   0.37%    49.77MiB / 990.7MiB    5.02%

NET I/O            BLOCK I/O           PIDS
1.26MB / 1.88MB    89.1GB / 1.42MB     15

21.8kB / 28.9kB    184GB / 15.5MB      22
```

It is of note that there are two containers running on your system and that docker stats does not give them human-readable names. You can intuit from the memory usage that 698ba3322462 is the Jupyter container. You can kill the docker stats tool with Ctrl+C and use docker ps to verify this, as in Listing 9-36.

Listing 9-36. Use docker ps to Diagnose a Memory Error

```
CONTAINER ID IMAGE                          ... NAMES
698ba3322462 ch9jupytermongo_this_jupyter ... ch9jupytermongo_this_jupyter_1
a37b17785360 mongo                          ... ch9jupytermongo_this_mongo_1
```

Sure enough, it is the Jupyter container that is using more than half of the system memory ... and in a failed state! It did not even finish loading the model. According the spaCy docs on their particular models,[10] it appears that loading the English model requires 1GB of RAM. Considering that this is the entirety of the RAM on your t2.micro, you will not be able to load the spaCy English model on a t2.micro.

Let's set the solution of this problem aside for a moment. First, you simply insert the tweet into MongoDB (Listing 9-37) as you did in Chapter 8. You do so by importing the pymongo module and instantiating a client to the database, before using that client to insert this_tweet. In Listing 9-38, you count the number of tweets in your tweet_collection to verify its insertion.

Listing 9-37. Insert a Single Tweet into MongoDB

```
In [6]: import pymongo

        mongo_cli = pymongo.MongoClient('this_mongo')
        result = (mongo_cli
                    .twitter_database
                    .tweet_collection
                    .insert_one(this_tweet) )
```

Listing 9-38. Count Tweets in tweet_collection

```
In [6]: (mongo_cli
            .twitter_database
            .tweets_collection.count())
Out[6]: 1
```

Scaling an AWS Application via Instance Type

In Chapter 1, you explored memory usage for various sized datasets, models, and model fitting procedures. The purpose of this was to examine memory constraints on an AWS t2.micro, the recommended system for working through this text. At the time, this examination was thoroughly academic. Here, you have hit an actual system constraint.

---

[10]https://spacy.io/docs/usage/models

You wish to load a language model available through the spaCy library that simply can't fit on your t2.micro. Using Docker, Docker-Compose, and AWS, you will create an efficient method for solving this problem. To do this, you

1. Shut down your Docker Compose application but keep your data volume.

2. Shut down your AWS instance.

3. Change the instance type of your AWS instance to a type that can meet your requirements, a t2.medium

4. Restart the AWS Instance, taking note of the new IP address generated.

5. Restart the Docker Compose application.

6. Perform your computation.

In Listing 9-39, you prepare to make the changes to your AWS instance by shutting down your application. Because you issue only a basic docker-compose down command, your Docker volume will persist through the entire process and you will not lose your MongoDB of tweets. It contains only one tweet, but this is sufficient for demonstration purposes.

Listing 9-39. Shut Down the Docker Compose Application

```
$ docker-compose down Stopping ch9jupytermongo_this_jupyter_1 ... done
Stopping ch9jupytermongo_this_mongo_1 ... done
Removing ch9jupytermongo_this_jupyter_1 ... done
Removing ch9jupytermongo_this_mongo_1 ... done
Removing network ch9jupytermongo_default
```

In Figure 9-5, you navigate to the EC2 control panel and shut down your running instance. To do this, you select "Stop" from the "Instance State" menu item on the Actions Menu. This is the first step in changing the instance type. The instance must be stopped in order to make the changes.

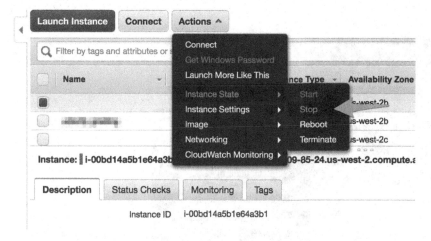

Figure 9-5. *Stop the AWS instance*

You have been working on a t2.micro. Figure 9-6 shows the on-demand instance pricing and technical specs for instances launched in the US West (Oregon) region. As is shown, AWS t2.micro has 1GB of RAM, insufficient to load the spaCy library you wish to use. To play it safe, change your instance type to a t2.medium, which has 4GB of RAM. This should be more than enough to load the 1GB spaCy model. In Figure 9-7, you change your AWS instance type. Select "Change Instance Type" from the "Instance Settings" menu item on the Actions menu.

Region: | US West (Oregon) | ♦

	vCPU	ECU	Memory (GiB)	Instance Storage (GB)	Linux/UNIX Usage
General Purpose - Current Generation					
t2.nano	1	Variable	0.5	EBS Only	$0.0059 per Hour
t2.micro	1	Variable	1	EBS Only	$0.012 per Hour
t2.small	1	Variable	2	EBS Only	$0.023 per Hour
t2.medium	2	Variable	4	EBS Only	$0.047 per Hour
t2.large	2	Variable	8	EBS Only	$0.094 per Hour
t2.xlarge	4	Variable	16	EBS Only	$0.188 per Hour
t2.2xlarge	8	Variable	32	EBS Only	$0.376 per Hour

Figure 9-6. *On-demand instance pricing and specs for US West (Oregon)*

Figure 9-7. *Change instance type*

Having changed the instance type, in Figure 9-8, you start the instance again using "Start" from the "Instance State" menu item in the Actions menu.

Figure 9-8. *Start the AWS instance*

■ **Warning**   Stopping your AWS instance will release the IP address you have been using. When the instance is started again, you will need to obtain the new IP address assigned to your system.

In Figure 9-9, you identify the new IP address assigned to your instance.

Figure 9-9. *The new IP address*

Restart Docker Compose Application

You have successfully modified the virtual hardware associated with your EC2 instance. You now restart your Docker Compose application. Because of your use of a Docker volume to persist your data, you should have no data loss during the process. In Listing 9-40, you reconnect to your AWS instance and navigate to the application directory.

Listing 9-40. Reconnect to the AWS Instance and Navigate to the Application Directory

```
(local) $ ssh ubuntu@ 54.244.99.222
  (AWS) $ cd ch_9_jupyter_mongo/
```

In Listing 9-41, you start the application in detached mode.

Listing 9-41. Start the Docker Compose Application in Detached Mode

```
$ docker-compose up -d
```

In Listing 9-42, you obtain the new security token for your Jupyter Notebook Server.

Listing 9-42. Obtain Jupyter Notebook Server Security Token

```
$ docker exec ch9jupytermongo_this_jupyter_1 jupyter notebook list
Currently running servers:
http://localhost:8888/?token=981014c28f5c8f694fd0321f418fddce6904f46857ace0
bc :: /home/jovyan
```

Complete the Computation

Having modified your system's virtual hardware, you return to the task at hand. In Listing 9-43, you walk through the steps of your work.

Listing 9-43. Rerun Preliminary Work

```
In [1]: from os import environ
        from twitter import OAuth

        oauth = OAuth(environ['ACCESS_TOKEN'],
                      environ['ACCESS_SECRET'],
                      environ['CONSUMER_KEY'],
                      environ['CONSUMER_SECRET'])

In [2]: from twitter import TwitterStream

        los_angeles_bbox = "-118.55, 33.97, -118.44, 34.05"
        twitterator = (TwitterStream(auth=oauth)
                       .statuses
                       .filter(locations=los_angeles_bbox))

In [3]: tw = next(twitterator)

In [4]: tw.keys()
Out[4]: dict_keys(['text', 'source', 'in_reply_to_status_id_str',
'favorited',
        'is_quote_status', 'in_reply_to_status_id', 'lang', 'filter_level',
        'geo',
        'favorite_count', 'created_at', 'entities', 'coordinates',
        'in_reply_to_user_id_str',
        'retweeted', 'truncated', 'retweet_count', 'id', 'contributors',
        'in_reply_to_user_id',
        'user', 'place', 'in_reply_to_screen_name', 'id_str',
        'timestamp_ms'])

In [5]: import pymongo

        mongo_server = pymongo.MongoClient('this_mongo')

In [6]: mongo_server.twitter.tweets.count()
Out[6]: 1

In [7]: result = (mongo_server
                  .twitter
                  .tweets
                  .insert_one(tw))

In [8]: mongo_server.twitter.tweets.count()
Out[8]: 2
```

Note that in Listing 9-43, Out[6], you receive an output of 1 for the count of tweets stored in your MongoDB. This verifies that the tweet you inserted into your MongoDB prior to changing your instance type has persisted through the change.

Encode Tweets as Document Vectors

In Chapter 8, you looked at serializing numpy vectors for insertion into both Redis and PostgreSQL. Here, you add an additional step to the process, before inserting a serialized numpy vector into MongoDB. Previously, these vectors were merely demonstration vectors and had no meaning. Now, you use the spacy.nlp English model to encode the tweets you have collected as numpy vectors representing the tweets, that is, as document vectors. In Listing 9-44, you load the spacy.nlp English model. This time the load is successful.

■ **Note**   I am purposefully avoiding an in-depth discussion of the spaCy library because it's beyond the scope of this text. Readers interested in its use are referred to the library's documentation at http://spacy.io.

Listing 9-44. Import spacy and Load the English Model

```
In [9]: import spacy

        nlp = spacy.load('en')
```

In Listing 9-45, you perform a search of all tweet documents. The search returns not the results themselves, but a cursor you can use to iterate through the documents one by one. This will be useful, as you will never have more than one document in memory at a time. You use the .next() class function to retrieve the first tweet.

Listing 9-45. Create a Cursor for a Search of All Tweets and Retrieve a Single Tweet

```
In [10]: cursor = mongo_server.twitter.tweets.find()
         stored_tweet = cursor.next()
```

In Listing 9-46, you display the text of the tweet.

Listing 9-46. Display Text of a Single Tweet

```
In [11]: stored_tweet['text']
Out[11]: 'Amazing day exploring the Sunken City.  #California is
unbelievable....
         https://t.co/INlOo1znc7'
```

In Listing 9-47, you use the spacy.nlp English model to create a document object using the text from the tweet. This document object contains the associated document vector as the attribute .vector. In Listing 9-48, you display the dimension of this vector using the .shape attribute.

Listing 9-47. Create Document Vector from Tweet Text

```
In [12]: doc = nlp(stored_tweet['text'])
```

Listing 9-48. Display Shape of Document Vector

```
In [13]: doc.vector.shape
Out[13]: (300,)
```

In Listing 9-49, you update your `tweet` document in MongoDB. The `.update_one()` function takes two dictionary arguments. The first argument is a dictionary used to search for the document you wish to update. Here, you specify that you wish to update a document with an `_id` matching your `stored_tweet`. The second argument specifies the value(s) you wish to update. In this case, you wish to set the key `'document_vector'` to the serialized value of your document vector. Note that you use the `.tostring()` function to serialize your document vector for storage.

Listing 9-49. Update MongoDB `tweet` Document with Serialized Document Vector

```
In [14]: mongo_server.twitter.tweets.update_one(
             {'_id': stored_tweet['_id']},
             {'$set': {'document_vector': doc.vector.tostring()}})
Out[14]: <pymongo.results.UpdateResult at 0x7f2796c6d750>
```

You then repeat the process for the second `tweet`. In Listing 9-50, you retrieve the second `tweet` and display its text. In Listing 9-51, you create the document vector. In Listing 9-52, you update the `tweet` document in MongoDB.

Listing 9-50. Retrieve Next Tweet and Display Text

```
In [15]: stored_tweet = cursor.next()
         stored_tweet['text']
Out[15]: "Idk what's so soothing about their fingers changing"
```

Listing 9-51. Create Document Vector from Tweet Text

```
In [16]: doc = nlp(stored_tweet['text'])
```

Listing 9-52. Update MongoDB `tweet` Document with Serialized Document Vector

```
In [17]: mongo_server.twitter.tweets.update_one(
             {'_id': stored_tweet['_id']},
             {'$set': {'document_vector': doc.vector.tostring()}})
Out[17]: <pymongo.results.UpdateResult at 0x7f2796c48fc0>
```

Switch AWS Instance Type to t2.micro

Having encoded the two tweets, you have completed the resource-intensive component of your task and no longer need the `spacy.nlp` English model. This means that you can switch your AWS instance type back to a `t2.micro`. This is done in the exact same fashion as switching to a `t2.medium`. Because you have written your results to MongoDB and used a Docker volume to persist data in MongoDB, your work will persist during the change. To make the change, you

1. Stop the Docker Compose application.

2. Shut down your AWS instance.

3. Change the instance type of your AWS instance back to a
    t2.micro.

4. Restart the AWS instance, taking note of the new IP address
    generated.

5. Restart the Docker Compose Application.

Retrieve Tweets from MongoDB and Compare

Using your t2.micro you can perform some comparisons of your tweet vectors. To do
this, you create a new Jupyter Notebook. In Listing 9-53, you connect to MongoDB.

Listing 9-53. Connnect to MongoDB

```
In [1]: import pymongo
        mongo_server = pymongo.MongoClient('this_mongo')
```

In Listing 9-54, you search all tweets in MongoDB using .find(). This time, you do
not work with a cursor; rather you cast the returned cursor to a list. Without going too
far into the Python of what you are doing, the effect is to create a list of tweets pulled
from MongoDB. Because you have used the keyword argument projection to request
document_vector, the list will contain only the _id and document_vector for each tweet.

Listing 9-54. Retrieve a List of Document Vectors from MongoDB

```
In [2]: tweet_vectors = list(mongo_server.twitter.tweets.
find(projection=['document_vector']))
```

In Listing 9-55, you deserialize the bytestream document vectors into numpy
vectors so that you can use them for a calculation. You do this, as you did in Chapter 8,
using the .fromstring() function.

Listing 9-55. Create a List of numpy Document Vectors

```
In [3]: import numpy as np
        tweet_vectors_np = [tw['document_vector'] for tw in tweet_vectors]
        tweet_vectors_np = [np.fromstring(tw) for tw in tweet_vectors_np]
```

In Listing 9-56, you perform a cosine similarity calculation between your two
tweet document vectors. The results show that these two tweets are very similar, at least
according to their spacy.nlp English model encoding.

Listing 9-56. Calculate Cosine Similarity of the Two Tweets

```
In [4]: from sklearn.metrics.pairwise import cosine_similarity
        cosine_similarity(tweet_vectors_np[0].reshape(1, -1),
                          tweet_vectors_np[1].reshape(1, -1))
Out[4]: array([[ 0.99992551]])
```

Docker Compose Networks

In Chapter 8, you launched and configured a PostgreSQL database on the same system as your Jupyter Notebook Server using a manually configured Docker Network. Although the process is non-trivial, working through that section can provide meaningful insights into how Docker configures networks internal to Docker to be used by containers to connect to each other. With Docker Compose there is an easier way.

■ **Note** Networking in Docker Compose is significantly different for docker-compose. yml files using Version 2 or higher. I continue to recommend the use of Version 3 and state this for completeness. This is to say that I will continue to operate as if you are working with Version 3 but you should be aware of the significant upgrades to networking in Docker Compose between Version 1 and Version 2.

At runtime, Docker Compose automatically sets up a single network for the application. Service containers defined in the docker-compose.yml file join this network by default, and are immediately available to other containers in the application and are discoverable by the name used to define the service. Consider the application defined in the sample docker-compose.yml file in Listing 9-57, presumed to be in a directory named ch_9_sample.

Listing 9-57. Sample docker-compose.yml File

```
version: "3"
services:
  this_jupyter:
    image: jupyter/scipy-notebook
    ports:
      - "8888:8888"
  this_redis:
    image: redis:alpine
  this_posstgres:
    image: postgres:alpine
```

When you run docker-compose up, docker-compose instructs the Docker Engine to

1. Create a network named ch9sample_default.

2. Create a container named ch9sample_this_jupyter with port 8888 in the container exposed over port 8888 on the host.

3. Instruct ch9sample_this_jupyter to join ch9sample_default using the name this_jupyter.

4. Create a container named ch9sample_this_redis.

5. Instruct ch9sample_this_redis to join ch9sample_default using the name this_redis.

6. Create a container named ch9sample_this_postgres.

7. Instruct ch9sample_this_ postgres to join ch9sample_ default using the name this_postgres.

Now each container in the application can access every other container using the container's name on the network, such as this_jupyter, this_redis, or this_postgres (Figure 9-10).

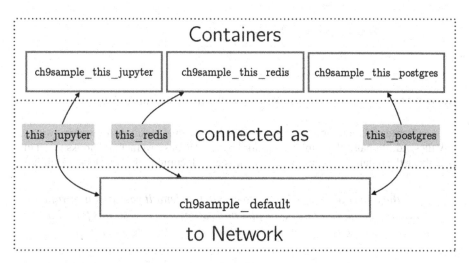

Figure 9-10. *A sample default Docker Compose network configuration*

Jupyter and Postgres with Persistence

For your final application in this chapter, you will build a Jupyter Notebook and PostgreSQL application (Figure 9-11). You will configure PostgreSQL to work with a data volume for persistence. You will also set up Postgres to use a build rather than an image.

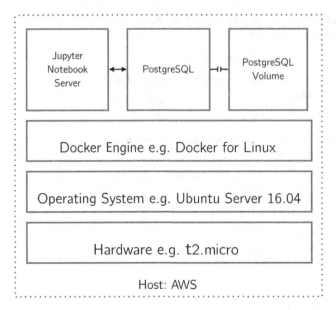

Figure 9-11. *A Docker Compose application with two services and a data volume*

In Chapter 8, I mentioned that PostgreSQL has a natural aptitude for working with CSV files. Additionally, the public image for PostgreSQL has several build hooks to aid in initializing the database at runtime. Per the postgres documentation on Docker Hub[11]:

> *After the entrypoint calls* initdb *to create the default postgres user and database, it will run any* *.sql *files and source any* *.sh *scripts found in that directory to do further initialization before starting the service.*

What this means is that you can add SQL files and shell scripts as part of the image build process that will execute automatically at runtime and set up your database for you.

Again, you begin by creating a directory to hold your project (Listing 9-58). In Listing 9-59, you create the new docker-compose.yml file (Listing 9-60).

Listing 9-58. Create a Directory to Hold the Project

```
$ mkdir ch_9_jupyter_postgres
$ cd ch_9_jupyter_postgres
```

Listing 9-59. Create docker-compose.yml File

```
$ vim docker-compose.yml
```

---

[11]https://hub.docker.com/_/postgres/

Listing 9-60. ch_9_jupyter_postgres/docker-compose.yml file

```
version: '3'
services:
  this_jupyter:
    build: docker/jupyter
    ports:
      - "8888:8888"
    volumes:
      - .:/home/jovyan
  this_postgres:
    build: docker/postgres
    volumes:
      - postgres_data:/var/lib/postgresql/data
volumes:
  postgres_data:
```

Specifying the Build Context

Readers will note that both services you define here use the build: keyword rather than the image: keyword. This means that your project will require two build contexts, one for each image. In Listing 9-61, you create two build contexts and then display your current project using tree.

Listing 9-61. Create Two Build Contexts

```
$ mkdir -p docker/jupyter
$ mkdir -p docker/postgres
$ tree
.
├────── docker
│   ├────── jupyter
│   └────── postgres
└────── docker-compose.yml
```

Next, in Listing 9-62, you create the first Dockerfile (Listing 9-63) for your this_jupyter service. You use this Dockerfile only to install the psycopg2 module you will be using to access PostgreSQL.

Listing 9-62. Create docker/jupyter/Dockerfile

```
$ vim docker/jupyter/Dockerfile
```

Listing 9-63. docker/jupyter/Dockerfile

```
FROM jupyter/scipy-notebook
USER root
RUN conda install --yes --name root psycopg2
USER jovyan
```

In Listing 9-64, you create the second Dockerfile (Listing 9-65) for your this_postgres service. In this Dockerfile, you use the postgres:alpine image as a base image and copy two files from the build context to the image, get_data.sh (Listing 9-66) and initdb.sql (Listing 9-67). As noted above, because you add these files to /docker-entrypoint-initdb.d/, at runtime the shell script will be executed by bash and the SQL file by PostgreSQL. You will use this to create a table in your database and populate it with data.

Listing 9-64. Create docker/postgres/Dockerfile

```
$ vim docker/postgres/Dockerfile
```

Listing 9-65. docker/postgres/Dockerfile

```
FROM postgres:alpine
COPY get_data.sh /docker-entrypoint-initdb.d/get_data.sh
COPY initdb.sql /docker-entrypoint-initdb.d/initdb.sql
```

■ **Note**   You obtain the CSV file you use to populate the database from the UCI Machine Repository.[12] There is an issue with this data in that some values are missing and have been replace with a '?' character. In order to deal with this, I have used the stream editor tool sed[13] to replace all instances of the '?' character with nothing (i.e. remove the '?' character from the file altogether).

Listing 9-66. docker/postgres/get_data.sh

```
#!/bin/bash
wget -P /tmp/ http://archive.ics.uci.edu/ml/machine-learning-databases/
breast-cancer-wisconsin/breast-cancer-wisconsin.data
sed 's/?//' /tmp/breast-cancer-wisconsin.data > /tmp/bcdata-clean.csv
```

Listing 9-67. docker/postgres/initdb.sql

```
CREATE TABLE bc_data (
    sample_id INTEGER UNIQUE PRIMARY KEY,
    clump_thickness INTEGER,
    uniformity_of_cell_size INTEGER,
    uniformity_of_cell_shape INTEGER,
    marginal_adhesion INTEGER,
    single_epithelial_cell_size INTEGER,
    bare_nuclei INTEGER,
    bland_chromatin INTEGER,
```

---

[12]http://archive.ics.uci.edu/ml/datasets.html
[13]www.gnu.org/software/sed/manual/sed.html

```
    normal_nucleoli INTEGER,
    mitoses INTEGER,
    class INTEGER
);
COPY bc_data FROM /tmp/bcdata-clean.csv DELIMITER ',' CSV;
```

In Listing 9-68, you use the tree tool to show the final state of your project.

Listing 9-68. Use tree to Show Application Directory

```
$ tree
.
├─── docker
│   ├─── jupyter
│   │   └─── Dockerfile
│   └─── postgres
│       ├─── Dockerfile
│       ├─── get_data.sh
│       └─── initdb.sql
└─── docker-compose.yml
```

Build and Run Your Application with Compose

In preparation for running your application, you use docker-compose build to build the two images used to define your services (Listing 9-69).

Listing 9-69. Build Application Using docker-compose build

```
$ docker-compose build
Building this_jupyter
Step 1/4 : FROM jupyter/scipy-notebook
...
Step 2/4 : USER root
...
Step 3/4 : RUN conda install --yes --name root psycopg2
...
Step 4/4 : USER jovyan
...
Successfully built b98e9ab6ee7e
Successfully tagged ch9jupyterpostgres_this_jupyter:latest
Building this_postgres
Step 1/3 : FROM postgres:alpine
...
Step 2/3 : COPY get_data.sh /docker-entrypoint-initdb.d/get_data.sh
...
Step 3/3 : COPY initdb.sql /docker-entrypoint-initdb.d/initdb.sql
...
Successfully built 97b956a4da7a
Successfully tagged ch9jupyterpostgres_this_postgres:latest
```

Finally, in Listing 9-70, you run your application using docker-compose up. Note that both a default network and your volume for data persistence are created prior to creating and starting your service containers.

Listing 9-70. Start the Compose Application jupyter_postgres

```
$ docker-compose up -d
Creating network "ch9jupyterpostgres_default" with the default driver
Creating volume "ch9jupyterpostgres_postgres_data" with default driver
Creating ch9jupyterpostgres_this_jupyter_1
Creating ch9jupyterpostgres_this_postgres_1
Starting ch9jupyterpostgres_this_jupyter_1
Starting ch9jupyterpostgres_this_postgres_1
```

In Listing 9-71, you display process information for your application.

Listing 9-71. Display Containers for Current docker-compose.yml

```
$ docker-compose ps
Name                                  Command
------------------------------------------------------------
ch9jupyterpostgres_this_jupyter_1     tini -- start-notebook.sh
ch9jupyterpostgres_this_postgres_1    docker-entrypoint.sh postgres

State     Ports
------------------------------------------------------------------------
Up        0.0.0.0:8888->8888/tcp
Up        5432/tcp
```

In Listing 9-72, you use docker-compose logs to display the logs associated with the this_postgres service.

Listing 9-72. Display Logs for this_postgres

```
$ docker-compose logs this_postgres
Attaching to ch9jupyterpostgres_this_postgres_1
this_postgres_1 | The files belonging to this database system will be owned
by user "postgres".
...
this_postgres_1 | /usr/local/bin/docker-entrypoint.sh: running /docker-
entrypoint-initdb.d/get_data.sh
this_postgres_1 | Connecting to archive.ics.uci.edu (128.195.10.249:80)
this_postgres_1 | breast-cancer-wiscon 100%
|****************************| 19889   0:00:00 ETA
...
this_postgres_1 | /usr/local/bin/docker-entrypoint.sh:
running /docker-entrypoint-initdb.d/initdb.sql
this_postgres_1 | CREATE TABLE
this_postgres_1 | COPY 699
...
```

What you wish to see here is the successful collection and insertion of the data. You can also connect to the running service via docker exec (Listing 9-73) to verify that the correct number of rows were inserted. Here you use the word count tool named wc[14] to count the number of lines in the file you downloaded.

Listing 9-73. Connect to this_postgres via docker exec

```
$ docker exec -it ch9jupyterpostgres_this_postgres_1 bash
bash-4.3# wc -l /tmp/breast-cancer-wisconsin.data
699 /tmp/breast-cancer-wisconsin.data
```

Last, you connect to Jupyter to test some code. In Listing 9-74, you perform a simple count of the number of rows in your bc_data table.

Listing 9-74. Count the Number of Rows in the bc_data Table

```
In [1]: import psycopg2 as pg2
In [2]: con = pg2.connect(host='this_postgres', user='postgres',
database='postgres')
        cur = con.cursor()
        cur.execute("SELECT COUNT(*) FROM bc_data;")
        cur.fetchall()
Out[2]: [(699,)]
```

Note that you have made use of the network created for you by specifying your host as 'this_postgres'.

Summary

In this chapter, I introduced the Docker Compose tool. You then used all of the techniques and tools discussed so far to build multi-service data applications using this tool. You built a trivial Jupyter-Redis application. You built a more complicated Jupyter-MongoDB application and explored the configuration of data persistence using Docker Compose. While using your Jupyter-MongoDB application you learned how to switch the underlying virtual hardware of your application if running as an AWS instance. Finally, you built a Jupyter-PostgreSQL application. In building the Jupyter-PostgreSQL application, you saw how to use build hooks defined in the postgres Docker image to load data into a database at runtime.

Having completed this chapter, I hope you will be able to design your own simple multi-service applications using Jupyter and any or all of the data stores I have introduced. In the next chapter, I will revisit the interactive programming paradigm and introduce the idea of building software with this paradigm at its core. You will use Docker Compose to build this software.

---

[14]https://linux.die.net/man/1/wc

CHAPTER 10

■ ■ ■

Interactive Software Development

Developing software as a data scientist is different from traditional software engineering and far less understood. For the traditional software developer, For any language, framworks built around reuse, extensibility, and stability exist. The most famous of these might be the Rails framework for the Ruby language. Rails is written from the ground up around its adopted paradigm, the Model-View-Controller design pattern, a pattern heavily favored in the implementation of user-facing software. Listing 10-1 shows the creation of and the default file structure for a new Rails application. Note that the new application has clear directories created for it based upon the usage pattern.

Listing 10-1. A Default Rails Application

```
$ rails new myapp
      create
        ...
$ tree -L 1 myapp/app
myapp/app/
├── assets
├── channels
├── controllers
├── helpers
├── jobs

├── mailers
├── models
└── views
```

Data science-specific software development has no such design pattern around which a similar framework might be built. In Chapter 3, I introduced the idea of interactive computing as an alternative to conventional programming. In this chapter, I propose that the idea of interactive computing itself be adopted as the cornerstone idea for a potential framework. You'll develop a project framework with infrastructure defined by a docker-compose.yml, built around Jupyter as your interactive computing

© Joshua Cook 2017
J. Cook, *Docker for Data Science*, DOI 10.1007/978-1-4842-3012-1_10

driver. The goals of this framework are aligned with those of an interactive computing project. This framework should facilitate ease in

- Iteration

- Scaling and distribution of hardware

- Sharing and documentation of work

A Quick Guide to Organizing Computational Biology Projects

For inspiration for this framework, let's look at the work of William Noble of the University of Washington.[1] Noble's work describes "one good strategy for carrying out computational experiments," focusing on "relatively mundane issues such as organizing files directories and documenting progress."

Noble focuses on a few key principles to structuring a project:

- File and directory organization

- Documenting work

- Executing work

- Version control

Figure 10-1 shows Noble's diagram for file and directory organization for a sample project called msms.

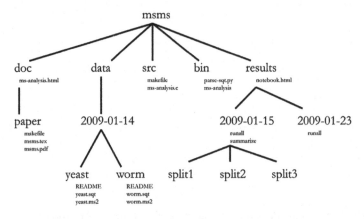

Figure 10-1. *Noble's sample project, msms*

---

[1]Noble, William Stafford; "A Quick Guide to Organizing Computational Biology Projects," PLOS Computational Biology, July 31, 2009, http://journals.plos.org/ploscompbiol/article?id=10.1371/journal.pcbi.1000424.

A Project Framework for Interactive Development

You'll draw directly upon this work to develop your framework. You'll use Jupyter Notebooks, numbered in sequence, as a method for both documenting and executing your work. These notebooks become a detailed record of activity as well as the means by which you drive this activity. Furthermore, you'll present a directory hierarchy designed around the use of the Jupyter Notebook as the driver of your work. Figure 10-2 shows a directory hierarchy built for interactive development.

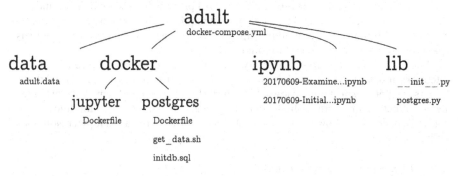

Figure 10-2. Directory hierachy built for interactive development

You'll build the directory hierarchy of your project using the following directories:

- data

  - Contains raw data files

- docker

  - Contains a subdirectory for each image to be defined using a build

  - Each subdirectory will become the build context for the respective image

- ipynb

  - Contains all Jupyter Notebook files

  - Replaces bin, doc, and results directories

  - Notebooks are drivers, scripts, documentation, and presentation

  - Notebooks are named with date and activity to sort them in place

- lib

  - Contains project-specific code modules, defined in the course of project development

Project Root Design Pattern

In Chapter 3, I proposed that

> *Jupyter doesn't replace vim, Sublime Text, or PyCharm. Jupyter replaces*
> *if __name__ == "__main__":.*

The if __name__ == "__main__": design pattern provides a launch hook for running a Python program. The project framework I propose here is not built around running code in such a way, and as such does not require such a launch hook. Rather, you are building this framework around the Jupyter Notebook as a driver. What you require is a pattern for importing modules into your notebooks.

Maintaining a clean project directory structure requires you to keep your notebooks and your Python modules in separate directories. Furthermore, I hold that it is less aesthetic to nest one inside of the other. This causes a problem at import time. Given a directory structure as shown in Listing 10-2, you can't import a module (e.g. some_module.py shown in Listing 10-3) from lib/ directly into a Jupyter Notebook module in ipynb/.

Listing 10-2. Sample Project Structure

```
$ tree
.
├── ipynb
│   └── some_notebook.ipynb
└── lib
    ├── __init__.py
    ├── some_module.py
```

Listing 10-3. A Demo Python Module, some_module.py

```
#!/bin/python

def say_hello ():
    print("Hello!")
```

Let's solve this problem by using what I will refer to as the **project root design pattern** (Listing 10-4).

Listing 10-4. The Project Root Design Pattern

```
In [1]: from os import chdir
        chdir('/home/jovyan')
```

The project root design pattern changes the current working directory of the Python kernel to be the root of the project. This is guaranteed by the configuration of the mounted volume in your docker-compose.yml file (Listing 10-8), where you mount the current directory (.) to the working directory of the jupyter image, home/jovyan. Thus, in running chdir('/home/jovyan') in a Jupyter Notebook (running on a jupyter image), you can guarantee that you will be at the project root. Furthermore, since you are using an absolute path in your chdir statement, you can run this command idempotently. Running this as the first command in any Jupyter Notebook means that you can import from your lib directory at will (Listing 10-5).

Listing 10-5. Import from lib.some_module.

```
In [2]: from lib.some_module import say_hello
        say_hello()

        Hello!
```

Initialize Project

In Chapter 9, you used a docker-compose.yml file to design an application consisting of a Jupyter Notebook Server and a PostgreSQL database. You used the docker-compose build tool and the design of the postgres image to gather your data and seed your database. Here, you do the same again, collecting your data from the UCI Machine Learning Repository. In this chapter, however, you formalize the process of gathering the data, documenting the process using a Jupyter Notebook.

In Listing 10-6, you initialize the project. You create a global directory for your project, ch10_adult. You create three subdirectories within this project, docker/, ipynb/, and lib/. You create a new __init__.py[2] file using the touch command. This has the effect of making the lib/ directory into a Python module. Finally, you initialize the project repository as a git repository using git init.

Listing 10-6. Initialize the ch10_adult Project

```
$ mkdir ch10_adult
$ cd ch10_adult/
$ mkdir docker ipynb lib
$ touch lib/__init__.py
$ git init
Initialized empty Git repository in /home/ubuntu/ch10_adult/.git/
```

In Listing 10-7, you create the docker-compose.yml file (Listing 10-8), which will define the infrastructure of your project. Note that you start simple. At this phase, you only have a single service, a Jupyter Notebook Server.

---

[2]http://mikegrouchy.com/blog/2012/05/be-pythonic-__init__py.html

Listing 10-7. Create the docker-compose.yml File

```
$ vi docker-compose.yml
```

Listing 10-8. docker-compose.yml

```
version: '3'
services:
  this_jupyter:
    image: jupyter/scipy-notebook
    ports:
      - "8888:8888"
    volumes:
      - .:/home/jovyan
```

Having defined your infrastructure, you bring the application online (Listing 10-9). Since you have used the `image:` keyword rather than the `build:` keyword to define the image used for your service, it is not necessary to perform a build prior to the launching of your application. After launch, you use the `docker-compose ps` tool to examine the running containers associated with your application (Listing 10-10). In order to access Jupyter through your browser, you will need to obtain a current authentication token (Listing 10-11).

Listing 10-9. Launch Initial System

```
$ docker-compose up -d
Starting ch10adult_this_jupyter_1
```

Listing 10-10. Examine Running Containers

```
$ docker-compose ps
Name                        Command                    State          Ports
-------------------------------------------------------------------------------
ch10adult_this_jupyter_1  tini -- start-notebook.sh  Up    0.0.0.0:8888->8888/tcp
```

Listing 10-11. Obtain Authentication Token

```
$ docker exec ch10adult_this_jupyter_1 jupyter notebook list
Currently running servers:
http://localhost:8888/?token=d6dc404c5e3c25ffd993579aeb06eeb2a801c4cbc75f
727e :: /home/jovyan
```

Examine Database Requirements

At this point, your application consists of a single service, a Jupyter Notebook server. The next phase in project development is to bring a database online. Here, you launch a notebook and use pandas[3] to examine a sample of the data in order to develop a schema to handle your data. From there you will prepare your `postgres` build context in order to seed your database.

---

[3]`http://pandas.pydata.org`

You begin by navigating to the Jupyter Notebook server in your browser. Note that the home directory of the Notebook server is comprised of the files of the root directory of your project.

In Figure 10-3, you navigate to the ipynb directory where you will create a new Python 3 Notebook (Figure 10-4).

Figure 10-3. *Navigate to the* ipynb *directory*

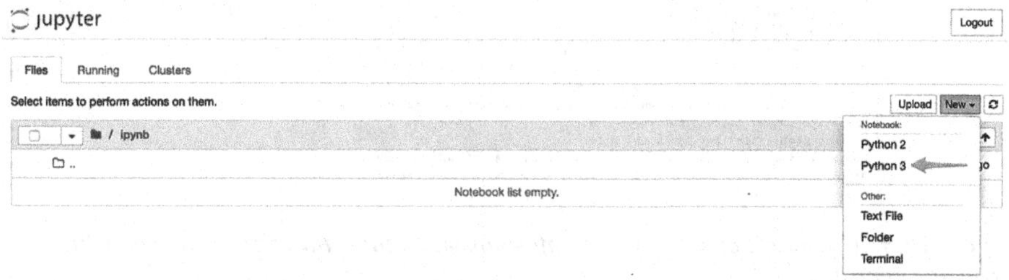

Figure 10-4. *Create a new Python 3 notebook*

Noble cites the need for "a chronologically organized lab notebook." I propose that in this project framework this need is met by the Jupyter Notebook. To chronologically organize your work, you simply name your notebook files with a year-month-date format followed by a high-level description of the task to be performed. So name this first notebook 20170611-Examine_Database_Requirements.ipynb. In Listing 10-12, you begin the notebook with the **project root design pattern**, after which you import pandas (Listing 10-13).

Listing 10-12. The Project Root Design Pattern

```
In [1]: from os import chdir
        chdir('/home/jovyan')
```

Listing 10-13. Import Necessary Libraries

```
In [2]: import random
        import pandas as pd
```

In Figure 10-5, you use Jupyter's capacity for annotation via markdown to include information on the dataset. The dataset is the `adult` dataset obtained from the UCI Machine Learning Repository.[4] You obtain the markdown you include from the dataset description included with the dataset.[5]

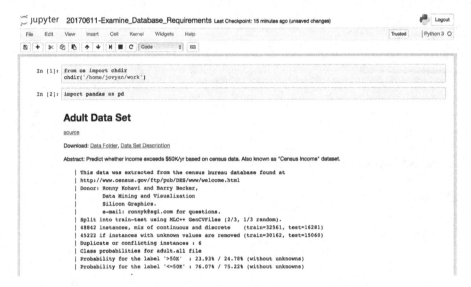

Figure 10-5. *Use markdown to include information on the dataset from UCI's Machine Learning Repository*

Listing 10-14 demonstrates the effect of the **project root design pattern** using a Jupyter shell call to `ls`. In Listing 10-15, you make another Jupyter shell call to `mkdir` to create a top-level directory `data`, after which you make a third Jupyter shell call to use the `wget` tool to obtain the dataset (Listing 10-16).

Listing 10-14. List Current Directory

```
In [3]: ls -l

        total 16
        drwxrwxr-x 2 jovyan 1000 4096 Jun 11 23:53 docker
        -rw-rw-r-- 1 jovyan 1000  145 Jun 11 23:55 docker-compose.yml
        drwxrwxr-x 3 jovyan 1000 4096 Jun 12 03:12 ipynb
        drwxrwxr-x 2 jovyan 1000 4096 Jun 11 23:53 lib
```

---

[4]http://archive.ics.uci.edu/ml/datasets/Adult
[5]http://archive.ics.uci.edu/ml/machine-learning-databases/adult/adult.names

Listing 10-15. Create Data Directory

```
In [4]: !mkdir data
```

Listing 10-16. Get the Dataset

```
In [5]: !wget -P data/ \
        http://archive.ics.uci.edu/ml/machine-learning-databases/adult/
        adult.data

        --2017-06-12 03:28:31-- http://archive.ics.uci.edu/ml/machine-learning-
        databases/adult/adult.data
        Resolving archive.ics.uci.edu (archive.ics.uci.edu)... 128.195.10.249
        Connecting to archive.ics.uci.edu (archive.ics.uci.edu)
        |128.195.10.249|:80... connected.
        HTTP request sent, awaiting response... 200 OK
        Length: 3974305 (3.8M) [text/plain]
        Saving to: 'data/adult.data'

        data/adult.data     100%[================>]   3.79M  7.93MB/s  in 0.5s

        2017-06-12 03:28:31 (7.93 MB/s) - 'data/adult.data' saved [3974305/3974305]
```

In Listing 10-17, you use the wc tool to count the number of lines in the file. In Listing 10-18, you use the head tool to see if the file has a header row. Note that the two rows shown are both data rows and therefore the file does not contain a header row.

Listing 10-17. Count the Number of Lines in the File

```
In [6]: !wc -l data/adult.data

        32562 data/adult.data
```

Listing 10-18. Check to See If the File Has a Header

```
In [7]: !head -n 2 data/adult.data

        39, State-gov, 77516, Bachelors, 13, Never-married, Adm-clerical,
        Not-in-family, White,
        Male, 2174, 0, 40, United-States, <=50K
        50, Self-emp-not-inc, 83311, Bachelors, 13, Married-civ-spouse,
        Exec-managerial,
        Husband, White, Male, 0, 0, 13, United-States, <=50K
```

In Listing 10-19, you load a 10% sample of the dataset using the pandas.read_csv function. To do this, you first define two variables, number_of_rows and sample_size. The first is the number of rows present in your dataset, the second the number of rows you would like to include in your sample. You next create a list, rows_to_skip, by using the random.sample function. random.sample takes two arguments: a list from which you wish to sample and a sample size. Here, you create a range of number from 0 to number_ of_rows and request a sample set of size number_of_rows - sample_size. The returned sample list is sorted and becomes your rows to skip. This list is passed to pandas.read_ csv at runtime. The effect is that pandas.read_csv will load a set of size sample_size.

The output of pandas.read_csv is a pandas.DataFrame object. You save this pandas. DataFrame as adult_df. You assign a list of column names (obtained from the dataset description at the UCI Machine Learning Repository) to the object attribute adult_ df.columns.

Listing 10-19. Load the File Using pandas

```
In [8]: number_of_rows = 32562
        sample_size = 3300

        rows_to_skip = random.sample(range(number_of_rows), number_of_rows -
        sample_size)
        rows_to_skip.sort()

        adult_df = pd.read_csv('data/adult.data', header=None, skiprows=rows_to_skip)
        adult_df.columns = [
            'age',
            'workclass',
            'fnlwgt',
            'education',
            'education_num',
            'marital_status',
            'occupation',
            'relationship',
            'race',
            'gender',
            'capital_gain',
            'capital_loss',
            'hours_per_week',
            'native_country',
            'income_label'
        ]
```

In Figure 10-6, you display a sample of the loaded DataFrame using adult_df.sample(3).

```
In [9]: adult_df.sample(3)
```

Out[9]:

	age	workclass	fnlwgt	education	education_num	marital_status	occupation	relationship	race	gender	capital_gain	capital_loss	hours_per_week	na
1571	43	Local-gov	36924	Masters	14	Married-civ-spouse	Prof-specialty	Husband	White	Male	0	0	40	l
2690	47	Private	456661	7th-8th	4	Married-civ-spouse	Craft-repair	Husband	White	Male	0	0	40	
2102	29	Private	203797	Some-college	10	Married-civ-spouse	Exec-managerial	Husband	Black	Male	0	0	40	l

Figure 10-6. *A sample of the loaded DataFrame*

At this point, you begin to construct a schema for loading your dataset into a PostgreSQL database. In Figure 10-7, you use Jupyter markdown to annotate the datatypes of the different feature columns as providing by UCI, and then examine the adult_df.dtypes to display the data types of each column in the pandas.DataFrame. If the data is properly formatted in the CSV file, then pandas will assign the proper data types to each column. If the pandas.DataFrame data types match the data types in the meta-information providing by UCI, then you should be able to count on the integrity of the data at load time, and be able to use the suggested data types to define your schema. Note that the pandas.DataFrame data types do match the data types in the meta-information and therefore you can use the meta-information as the basis for your schema without the need for special data handling.

Attribute Type

```
age: continuous
workclass: class
fnlwgt: continuous
education: class
education_num: continuous
marital_status: class
occupation: class
relationship: class
race: class
gender: class
capital_gain: continuous
capital_loss: continuous
hours_per_week: continuous
native_country: class
income_label: class
```

```
In [10]:   adult_df.dtypes

Out[10]:   age                int64
           workclass          object
           fnlwgt             int64
           education          object
           education_num      int64
           marital_status     object
           occupation         object
           relationship       object
           race               object
           gender             object
           capital_gain       int64
           capital_loss       int64
           hours_per_week     int64
           native_country     object
           income_label       object
           dtype: object
```

Figure 10-7. *Comparing meta-information from UCI's Machine Learning Repository to the* DataFrame *data types*

Having completed your preliminary work, you save the file and stop the notebook by selecting "Close and Halt" from the File Menu (Figure 10-8). You will use what you discovered here to build your database.

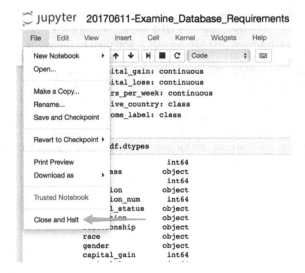

Figure 10-8. *Close and halt the notebook*

Managing the Project via Git

Before moving on to the next phase of the project, you commit all of your recent changes using git. In Listing 10-20, you check the status of your project (that is, what changes are present in your code) using the git status tool. Because you have not made any commits since you initialized the project, there will be changes associated with your initial docker-compose.yml file definition, in addition to the work you just did preparing to write your schema.

Listing 10-20. Check Project Status

```
$ git status
On branch master

Initial commit

Untracked files:
  (use "git add <file>..." to include in what will be committed)

        data/
        docker-compose.yml
        ipynb/
```

You will track each segment of work that you have done separately. First, in Listing 10-21, you add and commit the creation of your initial docker-compose.yml file. Then, in Listing 10-22, you do the same for your schema preparation work. Because this work is everything left unstaged you can use the -A flag to signify that you wish to add "all".

Listing 10-21. Add and Commit Initial docker-compose.yml File

```
$ git add docker-compose.yml
$ git commit -m 'initial docker-compose.yml file'
[master (root-commit) 3029c78] initial docker-compose.yml file
 1 file changed, 8 insertions(+)
 create mode 100644 docker-compose.yml
```

Listing 10-22. Add and Commit Schema Preparation Work

```
$ git add -A
$ git commit -m 'schema preparation work'
[master 13de8e1] schema preparation work
 3 files changed, 33336 insertions(+)
 create mode 100644 data/adult.data
 create mode 100644 ipynb/.ipynb_checkpoints/20170611-Examine_Database_
Requirements-checkpoint.ipynb
 create mode 100644 ipynb/20170611-Examine_Database_Requirements.ipynb
create mode 100644 docker-compose.yml
```

Note that this most recent commit also added an .ipynb_checkpoints directory. This is undesirable. In Listing 10-23, you create a .gitignore (Listing 10-24) file and add a few files you do not wish to track via git. In Listing 10-25, you reset the git HEAD to remove the most recent commit. Finally, in Listing 10-26, you perform the whole process once more.

Listing 10-23. Create the .gitignore file.

```
$ vi .gitignore
```

Listing 10-24. .gitignore

```
**/.ipynb_checkpoints
**/*.pyc
```

Listing 10-25. Reset git HEAD to the Penultimate Commit and Display Status

```
$ git reset HEAD~1
$ git status
On branch master
Untracked files:
  (use "git add <file>..." to include in what will be committed)

        .gitignore
        data/
        ipynb/

nothing added to commit but untracked files present (use "git add" to track)
```

Listing 10-26. Add and Commit Schema Preparation Work

```
$ git add -A
$ git commit -m 'schema preparation work'
[master ad896e2] schema preparation work
 3 files changed, 32951 insertions(+)
 create mode 100644 .gitignore
 create mode 100644 data/adult.data
 create mode 100644 ipynb/20170611-Examine_Database_Requirements.ipynb
```

■ **Warning** Be cautious when adding data files to a git commit. If syncing a local git repository with a cloud-based version-control system such as GitHub, the files must be less than 100MB to be uploaded to GitHub. Removing a file from a git commit, particularly when it was not the most recent commit, can be challenging and its presence will cause the sync to fail even if it has been removed from the most recent commit. The bash tool split[6] is recommended for breaking CSV files into smaller files that are less than 100MB if a cloud backup is desired.

Adding a Database to Your Application

You initially launched your application with a basic Jupyter service for some preliminary analysis of your dataset. Having done this, let's use what you learned to seed a PostgreSQL database with this dataset using an appropriate schema. Figure 10-9 shows a diagram of the next iteration of your system.

---

[6]https://ss64.com/bash/split.html

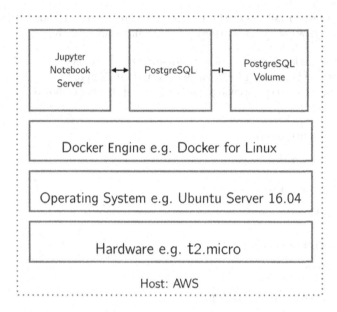

Figure 10-9. Second iteration of your application

To do this, you will add a few things to your docker-compose.yml file (Listing 10-27):

- Define the Jupyter Notebook service with a built image rather than an existing image in order to include your database interface library, psycopg2.

- Define a PostgreSQL service making use of the build hooks you saw in Chapter 9.

- Create a data volume to be used by the PostgreSQL service.

Listing 10-27. Next Version of Your docker-compose.yml

```
version: '3'
services:
  this_jupyter:
    build: docker/jupyter
    ports:
      - "8888:8888"
    volumes:
      - .:/home/jovyan
  this_postgres:
    build: docker/postgres
    volumes:
      - postgres_data:/var/lib/postgresql/data
volumes:
  postgres_data:
```

Here, you make use of several patterns you have seen before, but in particular you must take care to mount the new data volume to the correct location within the this_postgres container.

Next, you create directories for your two new build contexts (Listing 10-28) and create (Listing 10-29) a Dockerfile (Listing 10-30) to define your this_jupyter service. At this time, you make only a single change: adding psycopg2 to your Python 3 environment.

Listing 10-28. Create New Build Contexts

```
$ mkdir docker/jupyter docker/postgres
```

Listing 10-29. Create docker/jupyter/Dockerfile

```
$ vim docker/jupyter/Dockerfile
```

Listing 10-30. docker/jupyter/Dockerfile

```
FROM jupyter/scipy-notebook
USER root
RUN conda install --yes --name root psycopg2
USER jovyan
```

Then, you define the build context for your this_postgres service. As in Chapter 9, the build context for your PostgreSQL is much more involved and includes several files: a Dockerfile (Listing 10-31 and 32); a bash script, get_data.sh (Listing 10-34), to obtain and clean your data; and a sql file, initdb.sql (Listing 10-36), to initialize your database.

■ **Note** The CSV file for your data obtained from the UCI Machine Learning Repository includes a blank line at the end of the file. PostgreSQL is intolerant of any aberrant behavior and rejects the data copy in your docker/postgres/initdb.sql file (Listing 10-35) without special handling of this blank line. As in Chapter 9, I used the sed tool to handle this issue in the get_data.sh file (Listing 10-33). Here, I use the pattern /^\s*$/d. This has the effect of matching any lines comprised only of white space, including blank lines (^\s*$), and deleting them using the d command.

Listing 10-31. Create docker/postgres/Dockerfile

```
$ vim docker/postgres/Dockerfile
```

Listing 10-32. docker/postgres/Dockerfile

```
FROM postgres:alpine
COPY get_data.sh /docker-entrypoint-initdb.d/get_data.sh
COPY initdb.sql /docker-entrypoint-initdb.d/initdb.sql
```

Listing 10-33. Create docker/postgres/get_data.sh

```
$ vim docker/postgres/get_data.sh
```

Listing 10-34. docker/postgres/get_data.sh

```
#!/bin/bash
wget -P /tmp/ http://archive.ics.uci.edu/ml/machine-learning-databases/adult/adult.data
sed '/^\s*$/d' /tmp/adult.data > /tmp/adult-clean.csv
```

Listing 10-35. Create docker/postgres/Dockerfile

```
$ vim docker/postgres/initdb.sql
```

Listing 10-36. docker/postgres/initdb.sql

```
CREATE TABLE adult (
    age INTEGER,
    workclass TEXT,
    fnlwgt INTEGER,
    education TEXT,
    education_num INTEGER,
    marital_status TEXT,
    occupation TEXT,
    relationship TEXT,
    race TEXT,
    gender TEXT,
    capital_gain INTEGER,
    capital_loss INTEGER,
    hours_per_week INTEGER,
    native_country TEXT,
    income_label TEXT
);
COPY adult FROM '/tmp/adult.data' DELIMITER ',' CSV;
```

In Listing 10-37, you display your project using tree.

Listing 10-37. Display project

```
$ tree
.
├── data
│   └── adult.data
├── docker
│   ├── jupyter
│   │   └── Dockerfile
│   └── postgres
│       ├── Dockerfile
│       ├── get_data.sh
```

```
        └── initdb.sql
├── docker-compose.yml
├── ipynb
│   └── 20170611-Examine_Database_Requirements.ipynb
└── lib
```

In Listing 10-38, you launch your updated application. Note that you have added the --build flag to your docker-compose up command to ensure that the new build contexts are built before use. You then examine your running containers using docker-compose ps (Listing 10-39).

Listing 10-38. Launch the Application

```
$ docker-compose up -d --build
Creating network "ch10adult_default" with the default driver
Creating volume "ch10adult_postgres_data" with default driver
Building this_jupyter
...
Building this_postgres
...
Successfully built de96ba591ad9
Successfully tagged ch10adult_this_postgres:latest
Creating ch10adult_this_jupyter_1
Creating ch10adult_this_postgres_1
```

Configuring the seeding of a PostgreSQL database can be finicky. The difficulty of this task can be compounded when it is being done through a layer of abstraction, as you are doing here. Your initdb.sql file is being executed by the this_postgres container at runtime. This can make diagnosis and troubleshooting of any issues a challenge. I offer the following methods for verifying the success of database initialization:

- Confirm via docker-compose ps that the database is running. A syntax error in the initdb.sql file will cause the database to fail and exit at runtime. If this has happened, docker-compose ps will show a state of Exit for the ch10adult_this_postgres_1 container (Listing 10-39).

- Examine the logs for the this_postgres service (Listing 10-40). If your initialization was successful, the logs should contain a record of the execution of your initialization scripts and sql files. Note that in Listing 10-40, you can see that 32561 records have been successfully copied into the database.

- Connect to the running container via a docker exec call to the psql tool (Listing 10-41).

Listing 10-39. Examine Running Containers

```
$ docker-compose ps
            Name                    Command              State          Ports
--------------------------------------------------------------------------------
ch10adult_this_jupyter_1   tini--start-notebook.sh       Up      0.0.0.0:8888->8888/tcp
ch10adult_this_postgres_1  docker-entrypoint.sh postgres Up               5432/tcp
```

Listing 10-40. Examine this_postgres Logs

```
...
this_postgres_1  | /usr/local/bin/docker-entrypoint.sh: running /docker-
entrypoint-initdb.d/get_data.sh
this_postgres_1  | Connecting to archive.ics.uci.edu (128.195.10.249:80)
this_postgres_1  | adult.data               9%
|**                          |   363k  0:00:09 ETA
this_postgres_1  | adult.data              100%
|*****************************|  3881k  0:00:00 ETA
this_postgres_1  |
this_postgres_1  |
this_postgres_1  | /usr/local/bin/docker-entrypoint.sh: running /docker-
entrypoint-initdb.d/initdb.sql
this_postgres_1  | CREATE TABLE
this_postgres_1  | COPY 32561
...
```

Listing 10-41. Connect to this_postgres via psql

```
$ docker exec -it ch10adult_this_postgres_1 psql postgres postgres
psql (9.6.3)
Type "help" for help.

postgres=# SELECT COUNT(*) FROM adult;
 count
-------
 32561
(1 row)
```

After verifying that the this_postgres service is properly configured, you commit these infrastructure changes to your git log (Listing 10-42). In Listing 10-43, you add all files and commit the changes.

Listing 10-42. Check Project Status

```
$ git status
On branch master
Changes not staged for commit:
  (use "git add <file>..." to update what will be committed)
  (use "git checkout -- <file>..." to discard changes in working directory)
```

```
    modified:    docker-compose.yml
```

```
Untracked files:
  (use "git add <file>..." to include in what will be committed)

        docker/
```

Listing 10-43. Add and Commit Changes

```
$ git add -A
$ git commit -m 'add postgres service with database seed'
[master 5a35f01] add postgres service with database seed
 5 files changed, 37 insertions(+), 1 deletion(-)
 create mode 100644 docker/jupyter/Dockerfile
 create mode 100644 docker/postgres/Dockerfile
 create mode 100644 docker/postgres/get_data.sh
 create mode 100644 docker/postgres/initdb.sql
```

Since you have stopped and relaunched your Jupyter Notebook server, you will need to obtain a new authentication token in order to access the server in the browser once more (Listing 10-44).

Listing 10-44. Obtain the Authentication Token

```
$ docker exec ch10adult_this_jupyter_1 jupyter notebook list
Currently running servers:
http://localhost:8888/?token=6ab886ef19e02fe8ac351d0c28d03a50ab13be69a69b4
6d7 :: /home/jovyan
```

Interactive Development

A major goal of this project to framework is to facilitate a new style of software development called *interactive development*. The interactive development of modules is as follows.

1. Use Jupyter to write code interactively in a notebook.

2. When a block of code gets too large or needs to be repeated, abstract this code into a function in Jupyter.

3. Test the performance of this new function in Jupyter.

4. Move this function to a module in your library of code.

5. Import the code for use as needed.

Let's demonstrate the process here with a simple method you will use often, a basic connection from Jupyter to your database. In this case, you will abstract the function into your library of code to adhere to the best practice of not repeating code. You begin by navigating to ipynb/ and creating a new file. You rename the file with today's date and what you will be doing (e.g. 20170613-Initial_Database_Connection.ipynb). In Listing 10-45, you begin the notebook with the project root design pattern, after which you import psycopg2(Listing 10-46).

Listing 10-45. The Project Root Design Pattern

```
In [1]: from os import chdir
        chdir('/home/jovyan')
```

Listing 10-46. Import Necessary Libraries

```
In [2]: import psycopg2 as pg2
```

In Listing 10-47, you connect to your database as you have done previously, instantiating a connection and a cursor from that connection. You make use of the network created for you by Docker Compose and refer to the PostgreSQL by its name on the network, this_postges (that is, the same name you have given to the PostgreSQL service). In Listing 10-48, you use the cursor to execute a query to the database, print the results of the query, and then close the connection.

Listing 10-47. Connect to postgres and Create a Cursor

```
In [3]: con = pg2.connect(host='this_postgres', user='postgres',
database='postgres')
        cur = con.cursor()
```

Listing 10-48. Query the Database and Close the Connection

```
In [4]: cur.execute("SELECT COUNT(*) FROM adult;")
        print(cur.fetchall())
        con.close()

        [(32561,)]
```

The code in Listing 10-47 is code that you will be using often. Although it is just two lines of code, it is worth abstracting into a function because you will be using it with frequency. In Figure 10-10, you write this function in a Jupyter cell, and then use a tab completion to display the function's docstring.

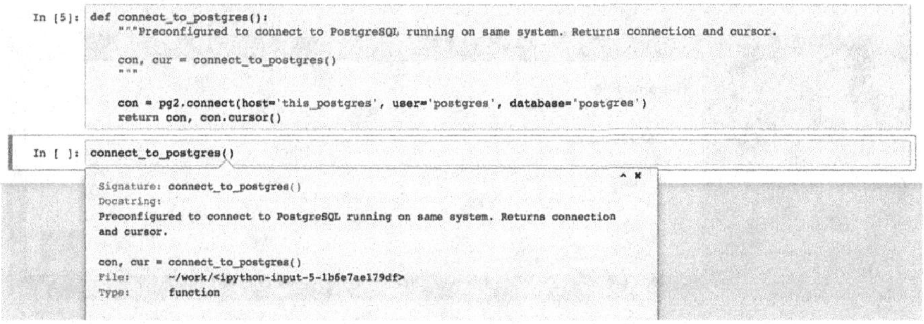

Figure 10-10. Define a function and display its docstring

You next verify that your connect_to_postgres function works as you expect in Listing 10-49.

Listing 10-49. Test connect_to_postgres

```
In [6]: con, cur = connect_to_postgres()
        cur.execute("SELECT COUNT(*) FROM adult;")
        print(cur.fetchall())
        con.close()

        [(32561,)]
```

Having verified that your function for accessing this_postgres works, you can add the function to an external Python module for import. Since you are done with this notebook, you should save and then close and halt the notebook.

Create a Python Module Using Jupyter

You will use the Jupyter Server's capacity for creating and editing text files to build a lib.postgres module. In Figure 10-11, you navigate to lib/ using the Notebook server, and then within lib/ you create a new text file. In Figure 10-12, you rename this file postgres.py. Next, you populate the new text file with the code in Listing 10-50.

Figure 10-11. *Create a new text file*

Figure 10-12. *Rename the next file to* postgres.py

234

Listing 10-50. `lib.postgres` Module

```
"""Helper module for interfacing with PostgreSQL."""
import psycopg2 as pg2

def connect_to_postgres():
    """Preconfigured to connect to PostgreSQL. Returns connection and cursor.

    con, cur = connect_to_postgres()
    """
    con = pg2.connect(host='this_postgres', user='postgres', database='postgres')
    return con, con.cursor()
```

Next, you create a new notebook to test the function you have written. It may be a bit pedantic to create a new notebook for each task. You do so here to highlight the desired workflow of the interactive development method. You create a new notebook titled 20170613-Verify_Database_Connection.ipynb. In Listing 10-51, you begin the notebook with the project root design pattern. In Listing 10-52, you import lib.postgres and verify that connect_to_postgres functions as you expect.

Listing 10-51. The Project Root Design Pattern

```
In [1]: from os import chdir
        chdir('/home/jovyan')
```

Listing 10-52. Test psql.connect_to_postgres

```
In [2]: import lib.postgres as psql
        con, cur = psql.connect_to_postgres()
        cur.execute("SELECT COUNT(*) FROM bc_data;")
        print(cur.fetchall())
        con.close()

        [(32561,)]
```

Finally, you track your work using git. In Listing 10-53, you check the status of your project. In Listing 10-54, you add and commit all of your recent work.

Listing 10-53. Check Status of Project

```
$ git status
On branch master
Untracked files:
  (use "git add <file>..." to include in what will be committed)

        ipynb/20170613-Initial_Database_Connection.ipynb
        ipynb/20170613-Verify_Database_Connection.ipynb
        lib/
```

Listing 10-54. Add All Files and Commit

```
$ git add -A
ubuntu@LOCAL:~/ch10_adult (master)
$ git commit -m 'initial database connection'
[master d2461f6] initial database connection
 4 files changed, 185 insertions(+)
 create mode 100644 ipynb/20170613-Initial_Database_Connection.ipynb
 create mode 100644 ipynb/20170613-Verify_Database_Connection.ipynb
 create mode 100644 lib/__init__.py
 create mode 100644 lib/postgres.py
```

In Listing 10-55, you display the current state of your project.

Listing 10-55. Current Project Status

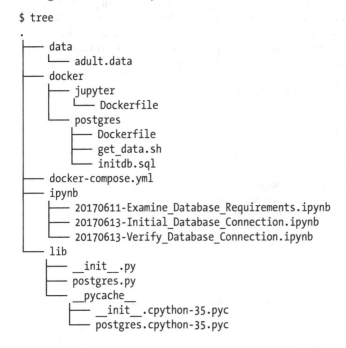

```
$ tree
.
├── data
│   └── adult.data
├── docker
│   ├── jupyter
│   │   └── Dockerfile
│   └── postgres
│       ├── Dockerfile
│       ├── get_data.sh
│       └── initdb.sql
├── docker-compose.yml
├── ipynb
│   ├── 20170611-Examine_Database_Requirements.ipynb
│   ├── 20170613-Initial_Database_Connection.ipynb
│   └── 20170613-Verify_Database_Connection.ipynb
└── lib
    ├── __init__.py
    ├── postgres.py
    └── __pycache__
        ├── __init__.cpython-35.pyc
        └── postgres.cpython-35.pyc
```

Add Delayed Processing to Your Application

You will now iterate on your application to its final state. You will add to your existing application a Redis service as well as two additional services defined as variations on the Jupyter image. In Chapter 9, I discussed how you might use Redis for caching intermediate results. Here you will explore another use for Redis as the backbone of a delayed job processing system. In addition to Redis, the delayed job processing system will use a Worker service, used for executing delayed jobs, and a Monitor service for monitoring the status of delayed jobs through a web browser. Figure 10-13 shows a diagram of your final application.

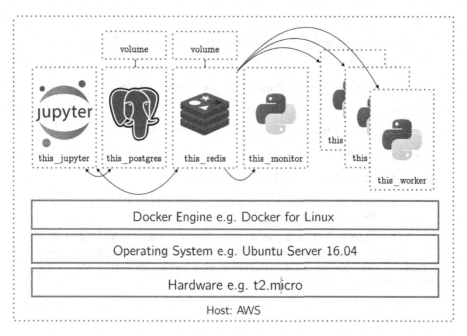

Figure 10-13. *Final application diagram*

To do this, you will add a few things to your docker-compose.yml file (Listing 10-56):

- Define a Redis service using the image: keyword.

- Create a data volume to be used by the Redis service.

- Define a Worker service using the same build context as the Jupyter service.

- Define a Monitor service using the same build context as the Jupyter service.

While both the Worker service and the Monitor service will be defined using the docker/jupyter build context, you will extend these images at runtime using the entrypoint: keyword. This keyword specifies the command with which the image should launch (in other words, the core process that will define the behavior of the container).

The Worker service will use the rqworker tool in order to interface with Redis to obtain and then execute queued jobs. You use the exec form of the entrypoint: keyword and take advantage of yaml lists to specify the instantiating process. The entrypoint consists of

- tini

  - The PID 1 tool mentioned in Chapter 5, used by Jupyter for instantiating all containers

237

- `--`
  - A best practice in instantiating a container with `tini`[7]
- `rqworker`
  - The process you will use to run your Worker service
- `-u`
  - The URL flag
- `redis://this_redis:6379`
  - The URL on which Redis will be available
  - Uses the Redis Service's name on the network created by Docker Compose

The Monitor service will use the `rq-dashboard` tool in order to provide a web-based dashboard for monitoring the status of queued jobs. You use the exec form of the `entrypoint:` keyword and take advantage of yaml lists to specify the instantiating process. The entrypoint consists of

- `tini`
- `--`
- `rq-dashboard`
  - The process you will use to run your Monitor service
- `-H`
  - The host flag
- `this_redis`
  - The Redis service's name on the network created by Docker Compose
- `-p`
  - The port flag
- `5000`
  - The port on which your Monitor service will be available

As before, you do not explicitly specify links between containers, letting Docker Compose establish the links for you. You make sure to connect the `redis_data` volume to the correct location within the `this_redis` container.

---

[7] `https://github.com/krallin/tini`

Listing 10-56. Next Version of Your docker-compose.yml

```yaml
version: '3'
services:
  this_jupyter:
    build: docker/jupyter
    ports:
      - "8888:8888"
    volumes:
      - .:/home/jovyan
  this_postgres:
    build: docker/postgres
    volumes:
      - postgres_data:/var/lib/postgresql/data
  this_redis:
    image: redis
    volumes:
      - redis_data:/data
  this_worker:
    build: docker/jupyter
    volumes:
      - .:/home/jovyan
    entrypoint:
      - "tini"
      - "--"
      - "rqworker"
      - "-u"
      - "redis://this_redis:6379"
  this_monitor:
    build: docker/jupyter
    volumes:
      - .:/home/jovyan
    ports:
      - "5000:5000"
    entrypoint:
      - "tini"
      - "--"
      - "rq-dashboard"
      - "-H"
      - "this_redis"
      - "-p"
      - "5000"
volumes:
  postgres_data:
  redis_data:
```

Next (Listing 10-57), you update the Jupyter Dockerfile (Listing 10-58) to include the necessary libraries to drive the Worker and Monitor services.

239

Listing 10-57. Update the docker/jupyter/Dockerfile

```
$ vim docker/jupyter/Dockerfile
```

Listing 10-58. docker/jupyter/Dockerfile

```
FROM jupyter/scipy-notebook
USER root
RUN conda install --yes --name root psycopg2
RUN conda install --yes --name root redis rq
RUN ["bash", "-c", "source activate root && pip install rq-dashboard"]
USER jovyan
```

In Listing 10-59, you display your project using tree.

Listing 10-59. Display Project

```
$ tree
.
├── data
│   └── adult.data
├── docker
│   ├── jupyter
│   │   └── Dockerfile
│   └── postgres
│       ├── Dockerfile
│       ├── get_data.sh
│       └── initdb.sql
├── docker-compose.yml
├── ipynb
│   ├── 20170611-Examine_Database_Requirements.ipynb
│   ├── 20170613-Initial_Database_Connection.ipynb
│   └── 20170613-Verify_Database_Connection.ipynb
└── lib
    ├── __init__.py
    ├── postgres.py
    └── __pycache__
        ├── __init__.cpython-35.pyc
        └── postgres.cpython-35.pyc
```

In Listing 10-60, you launch your updated application. Note that you have continued to use the --build flag with your docker-compose up command to ensure that the new build contexts are built before use. You then examine your running containers using docker-compose ps (Listing 10-61).

Listing 10-60. Launch Application

```
$ docker-compose up -d --build
Creating network "ch10adult_default" with the default driver
Creating volume "ch10adult_postgres_data" with default driver
Creating volume "ch10adult_redis_data" with default driver
...
Creating ch10adult_this_worker_1
Creating ch10adult_this_redis_1
Creating ch10adult_this_postgres_1
Creating ch10adult_this_jupyter_1
Creating ch10adult_this_monitor_1
```

Listing 10-61. Examine Running Containers

```
$ docker-compose ps
           Name                 Command              State        Ports
-------------------------------------------------------------------------------
ch10adult_this_jupyter_1  tini--start-notebook.sh    Up  0.0.0.0:8888->8888/tcp
ch10adult_this_monitor_1  tini--rq-dashboard -H th...  Up  0.0.0.0:5000->5000/tcp...
ch10adult_this_postgres_1 docker-entrypoint.sh postgres  Up              5432/tcp
ch10adult_this_redis_1    docker-entrypoint.sh redis ...  Up            6379/tcp
ch10adult_this_worker_1   tini--rqworker -u redis: ...  Up  8888/tcp Creating
```

Configuring the delayed job system can also be a challenge, just as in configuring the PostgreSQL database. Troubleshooting can be done using very similar methods.

- Confirm via docker-compose ps that the services are running.

- Examine the logs for the services (Listings 10-62 and 10-63).

- Connect to the running container via a docker exec call to the bash tool.

Listing 10-62. Examine this_monitor Logs

```
$ docker-compose logs this_monitor
Attaching to ch10adult_this_monitor_1
this_monitor_1  | RQ Dashboard version 0.3.8
this_monitor_1  |  * Running on http://0.0.0.0:5000/ (Press CTRL+C to quit)
```

Listing 10-63. Examine this_worker Logs

```
$ docker-compose logs this_worker
Attaching to ch10adult_this_worker_1
this_worker_1   | 20:28:59 RQ worker 'rq:worker:6a695d66b402.5' started,
                  version 0.6.0
this_worker_1   | 20:28:59 Cleaning registries for queue: default
this_worker_1   | 20:28:59
this_worker_1   | 20:28:59 *** Listening on default......
```

After verifying that the new services are properly configured, you commit these infrastructure changes to your git log (Listing 10-64). In Listing 10-65, you add all files and commit the changes.

Listing 10-64. Check Project Status

```
$ git status
On branch master
Changes not staged for commit:
  (use "git add <file>..." to update what will be committed)
  (use "git checkout -- <file>..." to discard changes in working directory)

        modified:   docker-compose.yml
        modified:   docker/jupyter/Dockerfile

no changes added to commit (use "git add" and/or "git commit -a")
```

Listing 10-65. Add and Commit Changes

```
$ git add -A
ubuntu@LOCAL:~/ch10_adult (master)
$ git commit -m 'add delayed job system'
[master 9564efd] add delayed job system
 2 files changed, 31 insertions(+), 1 deletion(-)
```

Since you have stopped and relaunched your Jupyter Notebook server, you will need to obtain a new authentication token in order to access the server in the browser once more (Listing 10-66).

Listing 10-66. Obtain Authentication Token

```
$ docker exec ch10adult_this_jupyter_1 jupyter notebook list
Currently running servers:
http://localhost:8888/?token=46e478574fe9cf238c6e2e6bc9b9daccb7efa7154dfd
9d08 :: /home/jovyan
```

Extending the Postgres Module

Let's once more demonstrate the interactive development paradigm described in this chapter as you extend the postgres module you previously created. This time you will first develop a function for encoding your target. This function will be executed row by row by one or more workers. You will

- Use Jupyter to write code interactively in a notebook.

- Abstract this code into a function in Jupyter.

- Test the performance of this new function in Jupyter.

- Move this function to a module in your library of code.

- Pass the function to a worker via the job queue.

You begin by navigating to ipynb/ and creating a new file. You rename the file with today's date and what you will be doing (e.g. 20170619-Develop_encoding_target_function.ipynb). In Listing 10-67, you begin the notebook with the project root design pattern, after which you import connect_to_ (Listing 10-68).

Listing 10-67. The Project Root Design Pattern

```
In [1]: from os import chdir
        chdir('/home/jovyan')
```

Listing 10-68. Import Database Connection

```
In [2]: from lib.postgres import connect_to_postgres
```

In Figure 10-14, you use a markdown cell to include the attribute type meta-information for your dataset, as you did in 20170611-Examine_Database_Requirements.ipynb.

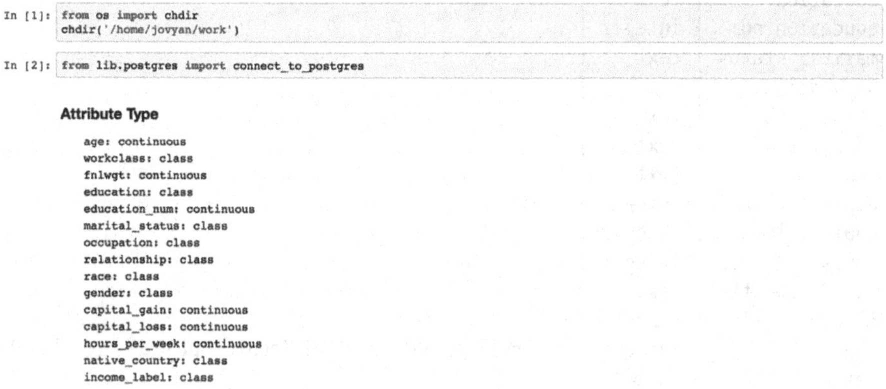

```
In [1]: from os import chdir
        chdir('/home/jovyan/work')

In [2]: from lib.postgres import connect_to_postgres
```

Attribute Type

```
age: continuous
workclass: class
fnlwgt: continuous
education: class
education_num: continuous
marital_status: class
occupation: class
relationship: class
race: class
gender: class
capital_gain: continuous
capital_loss: continuous
hours_per_week: continuous
native_country: class
income_label: class
```

Figure 10-14. *Display the attribute type meta-information*

In Listing 10-69, you create new columns in your database. As in Chapter 8, you manage your transactions manually via BEGIN and COMMIT statements. Note that you close the connection after each transaction.

Listing 10-69. Create New Columns

```
In [3]: con, cur = connect_to_postgres()
        cur.execute("""
        BEGIN;
        ALTER TABLE adult ADD COLUMN _id SERIAL PRIMARY KEY;
        ALTER TABLE adult ADD COLUMN target BOOLEAN;
        COMMIT;
        """)
        con.close()
```

It is a best practice to monitor changes to the database. In Listing 10-70, you use a psql via docker exec to examine the adult table.

Listing 10-70. Examine the adult Table via a docker exec psql Call

```
$ docker exec -it ch10adult_this_postgres_1 psql postgres postgres
psql (9.6.3)
Type "help" for help.

postgres=# \d adult
                        Table "public.adult"
      Column      | Type    |                  Modifiers
------------------+---------+--------------------------------------------------
-----
 age              | integer |
 workclass        | text    |
 fnlwgt           | integer |
 education        | text    |
 education_num    | integer |
 marital_status   | text    |
 occupation       | text    |
 relationship     | text    |
 race             | text    |
 gender           | text    |
 capital_gain     | integer |
 capital_loss     | integer |
 hours_per_week   | integer |
 native_country   | text    |
 income_label     | text    |
 _id              | integer | not null default nextval('adult__id_seq'::regclass)
 target           | boolean |
Indexes:
    "adult_pkey" PRIMARY KEY, btree (_id)
```

In Listing 10-71, you retrieve unique values for your target column, income_label.

Listing 10-71. Retrieve Unique Values for Target Column named income_label

```
In [4]: con, cur = connect_to_postgres()
        cur.execute("""SELECT DISTINCT(income_label) FROM adult;""")
        print(cur.fetchall())
        con.close()

        [(' >50K',), (' <=50K',)]
```

While `income_label` is categorical in nature, it only has two values and can thus be encoded as a Boolean value. In Listing 10-72, you write a short Jupyter script to do just this. You first query the database to retrieve the `_id` and `income_label` for a single row where the target column is NULL. You create a Boolean-valued variable `greater_than_50k`. Finally, you update the table for the given `_id` and close the connection to the database.

Listing 10-72. Encode a Single Instance's Target as a Boolean

```
In [5]: con, cur = connect_to_postgres()
        cur.execute("""SELECT _id, income_label FROM adult WHERE target IS NULL;""")
        this_id, income_label = cur.fetchone()

        greater_than_50k = (income_label == ' >50K')

        cur.execute("""
        BEGIN;
        UPDATE adult
        SET target = {}
        WHERE _id = {};
        COMMIT;
        """.format(greater_than_50k, this_id))

        con.close()
```

In Listing 10-73, you verify that the update was successful.

Listing 10-73. Verify Update

```
In [6]: con, cur = connect_to_postgres()
        cur.execute("""
        SELECT _id, income_label, target
        FROM adult WHERE _id = {};
        """.format(this_id))
        print(this_id, cur.fetchone())
        con.close()

        10 (10, ' >50K', True)
```

Having verified that your script works, you set about abstracting the script into a function (Listing 10-74).

Listing 10-74. `encode_target` Function

```
In [7]: def encode_target(_id):
            """Encode the target for a single row as a boolean value. Takes
            a row _id."""
            con, cur = connect_to_postgres()
```

```
cur.execute("""SELECT _id, income_label FROM adult where _id =
{}""".format(_id))
this_id, income_label = cur.fetchone()
assert this_id == _id

greater_than_50k = (income_label == ' >50K')

cur.execute("""
    BEGIN;
    UPDATE adult
    SET target = {}
    WHERE _id = {};
    COMMIT;
""".format(greater_than_50k, _id))

con.close()
```

In Listings 10-75 and 10-76, you test the new function and verify its success.

Listing 10-75. Select a New Row with Null Target and Encode

```
In [8]: con, cur = connect_to_postgres()
        cur.execute("""SELECT _id FROM adult WHERE target IS NULL;""")
        this_id, = cur.fetchone()
        encode_target(this_id)
        con.close()
```

Listing 10-76. Verify Encoding

```
In [6]: con, cur = connect_to_postgres()
        cur.execute("""
        SELECT _id, income_label, target
        FROM adult WHERE _id = {};
        """.format(this_id))
        print(this_id, cur.fetchone())
        con.close()

        11 (11, ' >50K', True)
```

Updating Your Python Module

In Figure 10-15, you navigate to lib/ using the notebook server, then within lib/ you select your postgres.py in order to update the module using the Jupyter Notebook server's text interface. Next, you add the code in Listing 10-74 to the file, as shown in Figure 10-16. Note that a check mark will appear next to the text file name when all current changes have been saved, as in Figure 10-17.

Figure 10-15. *Open postgres.py for editing*

```
 1  """Helper module for interfacing with PostgreSQL."""
 2  import psycopg2 as pg2
 3
 4  def connect_to_postgres():
 5      """Preconfigured to connect to PostgreSQL. Returns connection and cursor.
 6
 7      con, cur = connect_to_postgres()
 8      """
 9      con = pg2.connect(host='this_postgres', user='postgres', database='postgres')
10      return con, con.cursor()
11
12  def encode_target(_id):
13      """Encode the target for a single row as a boolean value. Takes a row _id."""
14      con, cur = connect_to_postgres()
15      cur.execute("""SELECT _id, income_label FROM adult where _id = {}""".format(_id))
16      this_id, income_label = cur.fetchone()
17      assert this_id == _id
18
19      greater_than_50k = (income_label == ' >50K')
20
21      cur.execute("""
22          BEGIN;
23          UPDATE adult
24          SET target = {}
25          WHERE _id = {};
26          COMMIT;
27      """.format(greater_than_50k, _id))
28
29      con.close()
```

Figure 10-16. *Latest version of postgres.py*

Figure 10-17. *All changes saved for postgres.py*

Next, you will create a new notebook to use the encode_target function via your delayed job system to encode all of the rows in the adult table. You create a new notebook titled 20170619-Encode_target.ipynb. In Listing 10-77, you begin the notebook with the project root design pattern. In Listing 10-78, you import the functions you need from lib.postgres.

Listing 10-77. The Project Root Design Pattern

```
In [1]: from os import chdir
        chdir('/home/jovyan')
```

Listing 10-78. Import Functions from lib.postgres

```
In [2]: from lib.postgres import connect_to_postgres, encode_target
```

In Listing 10-79, you see a new design pattern, the instantiation of a Queue from the rq library. The Queue is instantiated with a connection to a specific Redis server. As before, you use the name of your Redis service on the network created by Docker Compose, this_redis.

Listing 10-79. Create Connection to Redis and New Queue

```
In [3]: from redis import Redis
        from rq import Queue
        REDIS = Redis(host='this_redis')
        Q = Queue(connection=REDIS)
```

In Listing 10-80, you put all of the pieces together. You create a new connection to PostgreSQL. You use a for-loop to pull the row _id for 100 rows from the adult table where the target has not yet been encoded. For each row, you add the encode_target function with an associated _id to the job queue using the .enqueue() function. Note that the argument to be passed (that is, the _id, to encode_target at runtime) is passed as a second argument to .enqueue().

Listing 10-80. Add 100 Target Encoding Requests to Queue

```
In [4]: con, cur = connect_to_postgres()
        for _ in range(100):
            cur.execute("""SELECT _id FROM adult WHERE target IS NULL;""")
            this_id, = cur.fetchone()
            Q.enqueue(encode_target, this_id)
        con.close()
```

During at least one execution of this cell (you will need to run this particular cell over 300 times in order to encode the entire table unless modifications are made), I recommend using the browser-based monitor, as well as "tailing" the Docker Compose logs, to watch the Worker churn through these functions. The browser-based monitor will be available on the same IP address as your Jupyter Notebook server, but will be available on port 5000 (Figure 10-18). Listing 10-81 shows the "tailing" of the Docker Compose logs using the --follow flag. Each job process will pass through this log as it is executed.

Figure 10-18. *Browser-based Queue and Worker monitor*

Listing 10-81. Tailing the Docker Compose Logs

```
$ docker-compose logs --follow this_worker
...
this_worker_1   | 01:43:04 *** Listening on default...
this_worker_1   | 01:43:04 default: lib.postgres.encode_target(24)
                    (914a8229-a876-438f-98d4-d6cd39a469b2)
this_worker_1   | 01:43:04 default: Job OK (914a8229-a876-438f-98d4-d6cd39a469b2)
this_worker_1   | 01:43:04 Result is kept for 500 seconds
this_worker_1   | 01:43:04
this_worker_1   | 01:43:04 *** Listening on default...
this_worker_1   | 01:43:04 default: lib.postgres.encode_target(24)
                    (215ed343-0ca5-4f75-86e6-d8237e4a5983)
this_worker_1   | 01:43:04 default: Job OK (215ed343-0ca5-4f75-86e6-d8237e4a5983)
this_worker_1   | 01:43:04 Result is kept for 500 seconds
this_worker_1   | 01:43:04
this_worker_1   | 01:43:04 *** Listening on default...
this_worker_1   | 01:43:04 default: lib.postgres.encode_target(24)
                    (1dfe0282-cf90-49de-a904-8a2ced103c50)
this_worker_1   | 01:43:04 default: Job OK (1dfe0282-cf90-49de-a904-8a2ced103c50)
this_worker_1   | 01:43:04 Result is kept for 500 seconds
...
```

Finally, you track your work using git. In Listing 10-82, you check the status of your project. In Listing 10-83, you add and commit all of your recent work.

249

Listing 10-82. Check the Status of the Project

```
$ git status
On branch master
Changes not staged for commit:
  (use "git add <file>..." to update what will be committed)
  (use "git checkout -- <file>..." to discard changes in working directory)

        modified:   lib/postgres.py

Untracked files:
  (use "git add <file>..." to include in what will be committed)

        ipynb/20170619-Develop_encoding_target_function.ipynb
        ipynb/20170619-Encode_target.ipynb

no changes added to commit (use "git add" and/or "git commit -a")
lib/
```

Listing 10-83. Add All Files and Commit

```
$ git add -A
$ git commit -m 'function and queueing for target encoding'
[master 93f3033] function and queueing for target encoding
 3 files changed, 319 insertions(+)
 create mode 100644 ipynb/20170619-Develop_encoding_target_function.ipynb
 create mode 100644 ipynb/20170619-Encode_target.ipynb
```

In Listing 10-84, you display the current state of your project.

Listing 10-84. Current Project Status

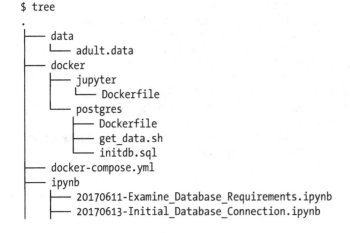

```
$ tree
.
├── data
│   └── adult.data
├── docker
│   ├── jupyter
│   │   └── Dockerfile
│   └── postgres
│       ├── Dockerfile
│       ├── get_data.sh
│       └── initdb.sql
├── docker-compose.yml
├── ipynb
│   ├── 20170611-Examine_Database_Requirements.ipynb
│   ├── 20170613-Initial_Database_Connection.ipynb
```

```
        ├── 20170613-Verify_Database_Connection.ipynb
        ├── 20170619-Develop_encoding_target_function.ipynb
        └── 20170619-Encode_target.ipynb
    └── lib
        ├── __init__.py
        ├── postgres.py
        └── __pycache__
            ├── __init__.cpython-35.pyc
            └── postgres.cpython-35.pyc
```

Summary

This chapter marks the conclusion of the book. In this chapter, you revisited the idea of interactive programming and saw the sketch of what a framework for interactive software development might look like. You defined the project root design pattern and software design pattern used to place Jupyter at the center of a well-structured interactive software application. You outlined steps for creating code modules via an interactive development process. Finally, you used Docker Compose and Redis to build a delayed job processing system into your application. Having finished this chapter, I hope that you are excited and prepared to begin building your own interactive applications.

Index

© Joshua Cook 2017

J. Cook, *Docker for Data Science*, DOI 10.1007/978-1-4842-3012-1

Get the eBook for only $5!

Why limit yourself?

With most of our titles available in both PDF and ePUB format, you can access your content wherever and however you wish—on your PC, phone, tablet, or reader.

Since you've purchased this print book, we are happy to offer you the eBook for just $5.

To learn more, go to http://www.apress.com/companion or contact support@apress.com.

Apress®

All Apress eBooks are subject to copyright. All rights are reserved by the Publisher, whether the whole or part of the material is concerned, specifically the rights of translation, reprinting, reuse of illustrations, recitation, broadcasting, reproduction on microfilms or in any other physical way, and transmission or information storage and retrieval, electronic adaptation, computer software, or by similar or dissimilar methodology now known or hereafter developed. Exempted from this legal reservation are brief excerpts in connection with reviews or scholarly analysis or material supplied specifically for the purpose of being entered and executed on a computer system, for exclusive use by the purchaser of the work. Duplication of this publication or parts thereof is permitted only under the provisions of the Copyright Law of the Publisher's location, in its current version, and permission for use must always be obtained from Springer. Permissions for use may be obtained through RightsLink at the Copyright Clearance Center. Violations are liable to prosecution under the respective Copyright Law.

Printed in the United States
by Bookmasters

Printed in the United States
By Bookmasters